Aids to Physiology

T. Scratcherd MB, BS (Dunelm), MD (Newcastle), FRCP (Edinburgh)

Formerly Professor of Physiology, University of Sheffield and Director of the Institute for Space Biomedicine, Sheffield

J. I. Gillespie BSc, PhD

Reader, Department of Physiological Science University of Newcastle Upon Tyne

THIRD EDITION

CHURCHILL
LIVINGSTONE

NEW YORK EDINBURGH LONDON MADRID MELBOURNE SAN FRANCISO TOKYO
1997

CHURCHILL LIVINGSTONE
Medical Division of Pearson Professional Limited

Distributed in the United States of America by Churchill
Livingstone Inc., 650 Avenue of the Americas, New York, N.Y.
10011, and by associated companies, branches and
representatives throughout the world.

First edition 1981
Second edition 1989
Third edition 1997

Standard edition ISBN 0 443 05451 7
International edition ISBN 0 443 05552 1

British Library of Cataloguing in Publication Data
A catalogue record for this book is available from the British
Library.

Library of Congress Cataloging in Publication Data
A catalog record for this book is available from the Library of
Congress.

Illustrated by Peter Cox

Medical knowledge is constantly changing. As new information
becomes available, changes in treatment, procedures, equipment
and the use of drugs become necessary. The authors and the
publishers have, as far as it is possible, taken care to ensure that
the information given in this text is accurate and up to date.
However, readers are strongly advised to confirm that the
information, especially with regard to drug usage, complies with
current legislation and standards of practice.

The
publisher's
policy is to use
**paper manufactured
from sustainable forests**

Printed in Hong Kong

Contents

Preface

In spite of the many revisions in the curricula of medical schools, a vast amount of knowledge is still expected from the student. This is particularly so in the pre-clinical stage when the student has so little time to digest all the information which is presented to him/her. This book is an attempt to help the hard-pressed student, by presenting most of the facts in note form to help in the revision process. It is an Aide-Mémoire and, because of the style in which it is written, may at first sight appear to the student as though he was reading someone else's notes. The student is advised to read an appropriate textbook thoroughly before turning to the *Aids* for revision purposes. It is important, when revising, also to study the tables and diagrams in the larger textbook carefully. Recall is the only effective way of learning and the *Aids* provides a list of headings which will help the student with minimum effort to test whether or not he can recall all the important facts. Students do not usually fail examinations because they do not understand but because they have omitted some important area.

Although primarily intended for medical students, other students studying elementary physiology courses should find it useful.

1997   T.S.
                                                             J.G.

Introduction

ADVICE TO STUDENTS

Many of the problems confronting students at the present time are a consequence of the vast amount of material presented to them to learn. Some of this has an immediate and obvious relevance to medicine and learning becomes easy. Other material does not appear to be relevant and students have to accept, as an act of faith, that what they are being taught is a necessary basis for medical practice and consequently is more difficult to learn. Perhaps the first advice which can be given is a warning. It is dangerous to think that important things come later. There is no special course which will teach you how to be a doctor.

Much work and effort can be saved by the adoption of efficient methods of study and students would do well to re-appraise their methods at the beginning of the medical course. There is a tendency to feel that the first weeks are a time which can be taken easily. Some schools have introduced continuous assessment to discourage this attitude and to take the strain off the student which previous intermittent examinations had imposed. Time is lost in other ways, e.g. failure to attend lectures and even when the student is conscientious in this respect, time may be lost by failure to concentrate and by day-dreaming. This can be prevented by strict attention to the lecture and by taking notes selectively. Do not attempt to take down all the lecturer says, only headings and essential information. After a lecture, read up the substance of the lecture in your main textbook and organize the knowledge in a systematic form using the *Aids* as a guide. There will be times when some concept has escaped you and you do not understand. Do not learn this by rote, turn to another textbook, it may help you, or discuss it with your student colleagues. A good way is to ask a demonstrator in the practical class or to raise it in a seminar. If that fails, then approach the lecturer directly.

It is assumed that you have a burning enthusiasm for medicine, there are no problems of motivation and that there is an intense desire to learn. If this is not so, then you should quickly consider if you are following the right course. Having attended a lecture and completed your note-taking and textbook study, including a careful study of figures and tables, the time has come to test how much you have learned. Without reference to the text, try to recall the salient points, making the briefest of notes. Then see what you have omitted by reference, either to your own notes or to the *Aids*. Repeat the process until all the material has been learned. Check on how much you have retained in a few days and then a few weeks later. If you have forgotten some of the material, repeat the process. The recall method of learning is an active process, it is hard but it is most effective. By using the *Aids* and your main textbook, revision should be carried out at odd moments such as between lectures, on the bus, as well as during your daily study period. By doing this, you will learn quickly and retain knowledge for long periods. It is important also to learn certain key line diagrams, such as the pressure curves of the cardiac cycle; block diagrams relating different events, such as iron metabolism; flow diagrams such as the renin–angiotensin–aldosterone system. These allow you to retain considerable information with relatively little effort.

1. The circulation

BLOOD

FUNCTION OF BLOOD

Blood is to be regarded as one of the largest organs of the body. It has the following functions:
- Respiratory gas transport.
- Transport of nutrients and waste products.
- Thermoregulation – distribution of heat from central organs to other parts of the body and skin.
- Haemostasis – to maintain a normal blood flow and to prevent blood loss following injury.
- Immunity – blood acts as a source and a transport system for immunocompetent cells and effector substances of the immune system.

VOLUME AND DISTRIBUTION OF BLOOD

	Systemic circulation (ml)	Pulmonary circulation (ml)
Arteries	550	400
Capillaries	300	60
Veins and venules	2150	840
Total	3000	1300

CONSTITUENTS OF BLOOD

Anticoagulated blood when centrifuged separates out into a column of fluid and cells comprising:
Plasma, a yellowish fluid, occupies about 55%.
Red cells occupy about 45%.
Haematocrit is the packed cell volume.
White cells settle as a buffy coat between the plasma and red cells.

PLASMA

About 3 L in adult or 4–5% body weight.
Male 40 ± 5.0 (SD) ml/kg.
Female 39 ± 5.0 (SD) ml/kg.

90% by weight is water, about 8% is protein, 2% organic compounds and electrolytes.

Plasma proteins
The plasma proteins are conventionally classified into albumin, globulin and fibrinogen. It is probably better to classify them according to their function.

Maintenance of the colloid osmotic pressure
Albumin.

Transport proteins
Haptoglobulin (α_2-globin). This complexes with haemoglobin liberated by vascular haemolysis. It is carried to the liver where the iron and protein are re-utilised.
Haemopexin has a similar function as haptoglobulin – both prevent the loss of iron from the body.
Caeruloplasmin – involved in the transport of copper.
Transferrin – a specific iron-binding protein-transporting iron.
Transcobalamins – transport vitamin B_{12}.
Albumin – a non-specific binder of many substances, e.g. bilirubin, folate, several hormones, fatty acids and haem.

Haemostatic proteins
Coagulation factors – as zymogens
Cofactors – V, VIII, XIII
Prothrombin
Fibrinogen.

Immunoglobulins
Most of the antibodies are gamma globulins made by plasma cells in lymphoid tissue. There are five different classes, designated IgA, IgD, IgE, IgG and IgM.

Other proteins
Complement – a family of proteins which is used in the induction of inflammation and immunity.
Proteolytic enzyme inhibitors:
    α_1-antitrypsin
    α_2-macroglobulin.
Lipoproteins.

THE RED CELL (erythrocyte)

Structure
Biconcave discs with an average diameter of about 7.2 μm, a volume of about 85 fl and a surface area of about 140 μm². Its shape maximises the surface area to volume ratio and thus facilitates gaseous exchange. The presence of a cytoskeleton allows it to be

readily deformable, allowing passage through narrow capillaries. The main constituent is the protein haemoglobin.

Numbers:
About 5×10^{12} cell per litre of blood
Reticulocytes 2% of mature cells

	Adult man	Adult woman
Red cell count	$4.5–6.5 \times 10^{12}$/L	$3.9–5.6 \times 10^{12}$/L
Packed cell volume	0.4–0.54	0.36–0.47
Haemoglobin	13.5–18 g/dl	11.5–16.4 g/dl

Absolute indices:

Mean corpuscular volume (MCV)	78–94 fl
Mean corpuscular haemoglobin (MCH)	27–32 pg
Mean corpuscular haemoglobin concentration (MCHC)	30–36 g/dl

MCV $= \dfrac{\text{packed cell volume fl } (10^{-15}\text{L})}{\text{red cell count}}$

e.g. $= 0.45/5 \times 10^{12} = 90 \times 10^{-15}\text{L}$

MCH $= \dfrac{\text{haemoglobin concentration} \times 10 \text{ pg } (10^{-12}\text{ g})}{\text{red cell count}}$

e.g. $= \dfrac{15 \text{ g/dl} \times 10}{5 \times 10^{12}} = 30 \times 10^{-12}$

MCHC $= \dfrac{\text{haemoglobin concentration g/dl}}{\text{packed cell volume (PCV)}}$

e.g. $= \dfrac{15 \text{ g/dl}}{0.45} = 33 \text{ g/dl}$

Red cell metabolism
In the process of maturation red cells lose their nucleus, mitochondria and ribosomes. Energy production is therefore dependent upon two pathways, anaerobic glycolysis to generate ATP and the hexose monophosphate pathway to protect against oxidative stresses.

Energy is required for two main reasons:
1. Maintenance of the intracellular cation balance – the high intracellular $[K^+]$ is maintained by an active sodium pump.
2. Maintenance of the characteristic shape and deformability of the red cell.

The hexose monophosphate pathway provides reducing power for three main reasons:
1. Prevention of membrane lipid oxidation.
2. Reduction of methaemoglobin – small amounts of methaemoglobin are formed within the cell by oxidation of the ferrous iron in haem to ferric, which is incapable of O_2 transport. It is essential to prevent the accumulation of methaemoglobin.
3. Detoxification of oxidants – in the production of methaemoglobin and during the ingestion of some antimalarial drugs, highly active oxygen radicals (superoxide) are produced

and this can lead to the production of H_2O_2. Detoxification occurs via the glutathione cycle.

Haemoglobin
Haemoglobin is a conjugated protein found in the red cell, responsible for the carriage of oxygen. It has a molecular weight of 64 500 daltons. It consists of four subunits, each a polypeptide chain, containing a molecule of haem. Haem is a protoporphyrin ring with a central atom of ferrous iron which combines reversibly with oxygen. Three haemoglobins are found in humans:

Haemoglobin A (HbA) – 98%
Haemoglobin B (HbB) – 2%
Haemoglobin F – present in the fetus and the newborn.

There are four types of globin chains which make up these haemoglobin molecules, α, β, δ and γ:

HbA – 2 α chains + 2 β chains
HbA$_2$ – 2 α chains + 2 δ chains
HbF – 2 α chains + 2 γ chains

Iron metabolism
A 70-kg man contains about 4 g iron which is distributed as follows:

Haemoglobin (circulating)	60%
Myoglobin	10%
Body stores	25%
Developing red cells	4%
Enzymes	1%
Plasma iron	0.1%

Storage and transport of iron
   Ferritin   is a ball-shaped protein which carries about 4500 iron atoms and the amount in plasma reflects the amount of stored iron.
   Haemosiderin   is formed of groups of ferritin molecules and is a store for iron, but is less metabolically active.
   Transferrin   has only one function, the transport of iron. This occurs from the mucosal cell to bone marrow (to red cell precursors), from iron stores in liver to other cells of the body and bone marrow (during growth and development), from the reticuloendothelial system (where iron is released from red cells) to stores and to cells when needed.

Iron balance
The body has three methods of maintaining balance:

1. The continuous use and re-use of iron from old red cells broken down in the reticuloendothelial system.
2. The regulation of iron absorption from the intestine. Iron

absorption is increased when iron is deficient in the body and decreased when there is sufficient so that overload is prevented.
3. The storage proteins release iron when there is excessive demand and store it at other times.

Daily requirements
There is no mechanism for iron excretion. Iron that is lost from the body comes about by desquamation of cells from the skin, into the gastrointestinal tract or urine, in tears and in sweat and by menstruation. There is also some loss during childbirth. Consequently requirements vary with age, sex and body weight:

• An adult male requires about 1 mg/day.
• An adult female during reproductive life requires about 2–3 mg/day. During menstruation, about 30–60 ml of blood will be lost (15–30 mg of iron).
• There is a high requirement in pregnancy for growth of the placenta and fetus, for the expansion of the maternal blood volume and to compensate for the loss which will occur at delivery.

The average intake of iron in the UK is about 14 mg/day, but only 10% of iron ingested is absorbed. The iron in food is in two forms, ferric iron or haem iron.

Dietary sources

There are two dietary sources, haem and non-haem iron, which can be found in:

Liver
Meat
Blood (black sausages)
Cereals, vegetables, pulses.

Iron absorption
Iron absorption is adjusted to demands and is determined by the ferritin content of the mucosal cells lining the gut. Transferrin is secreted from epithelial cells into the lumen and chelates the iron. The transferrin–iron complex binds to a receptor and enters the mucosal cell by a carrier-mediated transport mechanism and pinocytosis. Iron is also absorbed in the form of haem. Haem binds to receptors on the luminal membrane of the mucosa and is absorbed. Iron is released from the haem by haemoxygenase within the cell.

The absorbed iron forms a common pool in the mucosal cell. Iron which leaves the cells is complexed to transferrin in the interstitial space and is carried in this form by the blood. Iron not complexed in this way is removed from the portal blood by the hepatocytes.

Iron not released from the mucosal cell is sequestered in the mucosal cell as ferritin and is lost into the gut lumen and then from the body when the mucosal cell is shed from the apex of a villus. In the iron-replete stage, iron is incorporated into ferritin. In the iron-deficient state, iron passes into the transport pool and is complexed with transferrin.

Factors facilitating the absorption of iron are:

Low iron stores.
The state of activity of bone marrow.
Iron is absorbed in the ferrous state. This is brought about by the reducing properties of the gut content:

vitamin C
protein and amino acids
sugars
acid gastric juice.

Factors reducing the absorption of iron include:
High iron stores.
Phytic acid and phosphates in the diet.
A high pH in the gut.

Iron transport
Iron is transported as a complex with transferrin, a β-globulin with one molecule capable of binding two iron atoms. The iron-binding capacity of plasma is 300 μg/dl (one-third saturated). Ferritin (apoprotein plus iron) contains 26% iron when fully saturated, plasma levels reflect body iron stores. As it is an acute-phase protein its concentration may be increased in conditions such as rheumatoid arthritis and malignancy and will not reflect iron stores accurately. The iron content of a liver biopsy specimen is the best index of body stores.

Iron stores
Iron is stored as ferritin in reticuloendothelial cells of liver, spleen, bone marrow. It is also stored as haemosiderin, a conglomeration of ferritin molecules.

The origin of the erythrocyte

Embryonic
Mesoblastic stage (0–2 months) – from blood islands on the yolk sac.
Hepatic and splenic stage (2–5 months).
Myeloid stage (5 months until birth) – marrow of all bones.

Extrauterine
At birth – from the marrow of all bones.
5–7 years – erythropoiesis begins to disappear from centres of long bones.
Adult – erythropoiesis (red marrow) found only in flat bones, ribs, sternum, skull, pelvis, vertebrae and proximal ends of long bones.

Excess demand for red cells is met by spread of red marrow to infantile sites.

Erythropoiesis

There is small pool of multipotent cells in the bone marrow which are able to develop into lymphoid, myeloid and erythroid cells. The earliest recognisable erythroid precursor is called a burst-forming unit erythroid (BFUe), which passes through a series of stages of morphological and functional differentiation as erythroblasts and normoblasts before becoming a mature erythrocyte. This process is controlled by regulatory substances produced locally (from T-cells, monocytes and fibroid stromal cells) and outside the bone marrow (erythropoietin from the kidney). The single most important regulator of erythropoiesis is erythropoietin. The stages of development, the cell types and the regulatory factors are shown in Figure 1.1.

As cells go through this series of events (which takes a few days), they develop their typical morphological and functional differentiation, while their proliferative capacity is gradually lost. The nucleus gradually shrinks into a pyknotic mass. RNA appears in the cytoplasm. Synthesis of polypeptide chains and protoporphyrin occurs. Plasma transferrin binds to surface receptors and internalises by endocytosis to release iron. Haemoglobin is formed. The nucleus is extruded at the orthochromic normoblast stage and mitochondria disappear.

The red cell leaves the bone marrow as a reticulocyte (some RNA in cytoplasm), indicating immaturity. Reticulocytes contribute 0.5% of total red cell count, with the proportion increasing in response to conditions giving rise to rapid generation.

Life span of erythrocytes

Life span 120 days. About 25 g are removed per day and about 2.3×10^6 cells are produced every second. Erythrocyte survival is determined by marking erythrocytes and measuring the disappearance of marked cells from the circulation using:

- Transfusion of compatible but genetically different red cells.
- Labelling haem by incorporation of ^{15}N.
- Injecting erythrocytes labelled with ^{51}Cr.

The control of erythropoiesis

The fundamental stimulus for red cell production is hypoxia. Hypoxia is sensed by a receptor believed to be a haem-containing protein. Activation of the receptor stimulates the production of erythropoietin (a glycoprotein of molecular weight 28 000) from peritubular endothelial cells of the kidney. Erythropoietin is present in the blood and urine in cases of anaemia and is suppressed by overtransfusion of red cells. Erythropoietin stimulates maturation and development of erythrocyte precursor cells. Hypoxia is produced by a reduction in red cell mass or a change in environmental conditions such as ascent to altitude.

Stage	Cell type	Regulatory factors

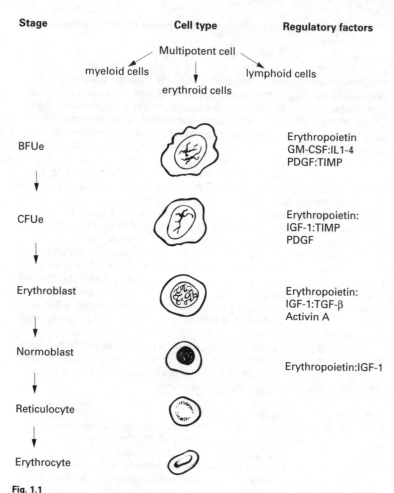

Multipotent cell

myeloid cells   lymphoid cells

erythroid cells

BFUe — Erythropoietin GM-CSF:IL1-4 PDGF:TIMP

CFUe — Erythropoietin: IGF-1:TIMP PDGF

Erythroblast — Erythropoietin: IGF-1:TGF-β Activin A

Normoblast — Erythropoietin:IGF-1

Reticulocyte

Erythrocyte

Fig. 1.1
Formation of blood cells. Activin A = a peptide; BFUe = burst-forming unit erythroid; CFUe = colony-forming unit erythroid; GM-CSF = granulocyte-macrophage colony stimulating factor; IGF = insulin-like growth factor; IL = interleukin; PDGF = platelet-derived growth factor; TIMP = tissue inhibitor of metalloproteinase; TGF-b = transforming growth factor b.

Substances essential to erythropoiesis
The following are required for erythropoiesis:

1. An adequate protein intake.
2. Vitamins:
   a. vitamin B_6 (pyridoxine) – required as a coenzyme in the

synthesis of aminolaevulic acid which is the first and rate-limiting step in the synthesis of haem;
b. vitamin B_{12} and folic acid (see section on vitamins, see p. 169) – required for DNA and RNA synthesis. In the absence of vitamin B_{12} or folic acid a megaloblastic anaemia occurs;
c. vitamin C – acts as a reducing agent and increases iron absorption and preserves folates during food preparation;
d. vitamin E – also acts as an antioxidant and prevents peroxidation of membrane lipids and red cell lysis.
3. Minerals:
a. iron;
b. copper – deficiency interferes with iron metabolism.
4. Other factors:
a. age and sex;
b. environment – at altitudes >2000 m there is hypoxic stimulation of erythropoiesis (one of the reasons for athletes training at high altitudes);
c. endocrines – thyroxine, cortisol, androgens and prolactin promote erythropoiesis.

Blood groups
Blood group antigens are found on the surface membrane of erythrocytes. Antibodies are found in the plasma. There are some 19 blood group systems, but only two are important routinely in the clinical situation. These are given the letters ABO and Rh (Rhesus).

The ABO system

Group	Antigen present on red cell	Antibody present in plasma
AB	A and B	—
A	A	Anti-B
B	B	Anti-A
O	—	Anti-A and anti-B

There are subgroups of the anti-A antigen. These are labelled A_1 and A_2. The practical importance of these is that A_2 and A_2B react more weakly with anti-A antibody than A_1 and A_1B cells, and A_2 and A_2B cells may react so weakly with anti-A antibodies that they can be wrongly grouped as being O and B respectively.

The Rhesus blood group system (Rh)
This system is clinically the most important after the ABO system because:
• The antigens are highly immunogenic.
• It is sometimes associated with haemolytic transfusion reactions.
• It is associated with haemolytic disease of the newborn.
There are five main antigens in this system labelled by the letters C, c, D, E and e. For practical purposes the Rh system is the product of three closely linked alleles, C and c, D and d, E and e, which are

inherited en bloc. This gives rise to eight haplotype combinations, CDE, CDe, CdE, Cde, cDE, cDe, cdE and cde. The d allele is thought to be amorphic and no red cell antigen has been described. Therefore, for this reason, the 'd antigen' can be thought of as an absence of antigen, and homozygotes are said to be *Rh negative*. Conversely, individuals who possess at least one D gene are said to be *Rh positive*.

Haemostasis

Haemostasis is the arrest of bleeding at a site of injury by the formation of a blood clot and depends upon a reaction both in the blood and in the vessel wall.

Blood coagulation – the formation of a clot
For descriptive purposes the process can be artificially divided into three pathways:

1. *An intrinsic pathway* factor X-activating system – follows contact of blood with subendothelial tissues or a foreign surface. All the components of the pathway are present in the blood.
2. *An extrinsic pathway* factor X activation – follows exposure of blood to damaged tissues providing tissue factor (thromboplastin).
3. *A common pathway* through which both extrinsic and intrinsic pathways are used to produce the fibrin clot.
   The diagrams below illustrate the three pathways.
   Note the following:
   a. The basic concept of the system is that of a cascade of successive zymogen activation.
   b. The extrinsic and intrinsic pathways allow for amplification of the process which is relatively slow.
   c. The common pathway is very fast and takes a matter of about 10 s, whereas the clotting time in a test tube is about 12 min. Control occurs in the extrinsic and intrinsic phases, without this, clotting beyond the site of damage might occur. Other factors which limit the coagulation to the site of damage are the binding of factors XII and XI to the damaged surface.
   d. The division into extrinsic and intrinsic pathways is artificial, but is useful for laboratory purposes.

The initiation of clotting by the intrinsic pathway
   Contact activation   Exposure of blood to a negatively charged surface causes adsorption of factor XII. Factor XII undergoes a conformational change which causes some pre-kallikrein to be converted into kallikrein. The kallikrein in the presence of HMWK (high-molecular-weight kininogen) changes XII to XII$_a$. Then, HMWK as cofactor XII$_a$ then converts XI to XI$_a$ (also bound to the charged surface).
   Pathways for the initiation of clotting by the intrinsic pathway and the stabilisation of the fibrin clot are shown below.

Common pathway

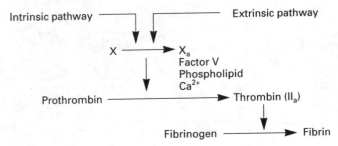

Intrinsic pathway ———┐ ┌——————— Extrinsic pathway

X ——————→ X_a
Factor V
Phospholipid
Ca^{2+}

Prothrombin ——————————————→ Thrombin (II_a)

Fibrinogen ——————→ Fibrin

Intrinsic pathway
Contact with a foreign surface

XII ——————→ XII_a

XI ——————→ XI_a

Ca^{2+}

IX ——————→ IX_a
Factor VIII
Phospholipid
Ca^{2+}

X ——————→ X_a

Extrinsic pathway

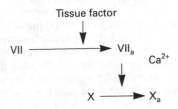

Tissue factor

VII ——————→ VII_a

Ca^{2+}

X ——————→ X_a

Stabilisation of the fibrin clot
The loose platelet plug is stabilised by the formation of a fibrin meshwork.

Role of the platelets and the formation of the primary haemostatic plug

Initiation of clotting

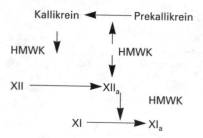

Stabilisation of the fibrin clot

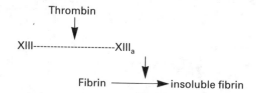

Platelets are involved in the formation of a platelet plug in the initial stages of the formation of a clot. The following sequence of events occurs:

Adhesion to a surface   Platelets may be activated by contact with subendothelial tissue, foreign or charged surfaces. Platelets adhere to and cover a damaged area by interaction with specific receptors.

Change of shape   A change in shape occurs after adhesion, from a discoid shape to one with an irregular outline and spiky projections.

Release of granule contents   Degranulation and release of contents follow. A-granules – contain substances which promote chemotaxis for fibroblasts and neutrophils; factors which promote adhesion; platelet–platelet interaction, and fibrinolytic inhibitor. Dense granules – ADP which promotes aggregation of platelets (ATP is a source of ADP); 5-hydroxytryptamine which promotes vasoconstriction.

Aggregation   Degranulation is accompanied by phosphatidyl inositol and arachidonic acid metabolism, both of which promote aggregation. The initial attachment of platelets to each other is by a bridge of fibrinogen, but ADP, calcium, collagen and thrombin are potent inducers of aggregation.

Consolidation   Clot retraction by platelet actin–myosin.

The role of prostaglandins in the formation of the haemostatic plug is shown below:

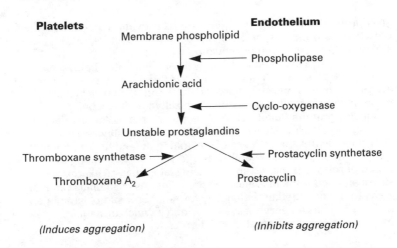

Platelets   **Endothelium**

Membrane phospholipid

Phospholipase

Arachidonic acid

Cyclo-oxygenase

Unstable prostaglandins

Thromboxane synthetase →   ← Prostacyclin synthetase

Thromboxane A_2 Prostacyclin

(Induces aggregation)   *(Inhibits aggregation)*

Inhibitors of blood coagulation
Some of the factors which keep the clot at the site of damage
have been described, but there are inhibitory mechanisms
which prevent the clotting process becoming out of control.
These are:
Anti-thrombin III
Heparin cofactor II
Activated protein C
α_2-Macroglobulin and α_1-antitrypsin.

Fibrinolysis
The major function of the fibrinolytic system is the degradation and
dissolution of formed fibrin within the circulation. The main enzyme
which degrades fibrin is plasmin, which is present in the circulation
in the inactive form plasminogen. Small amounts of free fibrin are
inactivated by irreversible binding to α_2-antiplasmin, the circulating

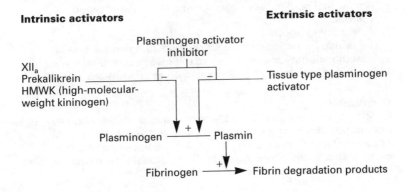

Intrinsic activators   **Extrinsic activators**

Plasminogen activator
inhibitor

XII_a
Prekallikrein
HMWK (high-molecular-
weight kininogen)

Tissue type plasminogen
activator

Plasminogen —— Plasmin

Fibrinogen ——→ Fibrin degradation products

glycoprotein. Plasminogen requires for its activation one of two groups of components, an extrinsic factor (the most important) and/or an intrinsic group of activation factors.

The pathways involved in fibrinolysis are illustrated below.

The control of the activation process is provided by plasminogen activator inhibitor, which is released from endothelial cells and which rapidly inactivates plasminogen activator. It has been suggested that the fluidity of blood in the microcirculation (where clotting more easily occurs because blood flow is slowest) is controlled by the slow release of plasminogen (and hence plasmin) from the endothelium of capillaries and venules. Plasminogen activator release is increased in certain states such as exercise, stress and under the influence of adrenaline. As tissue plasminogen activator binds together with newly formed fibrin and plasminogen, a blood clot has an enzyme system within it which provides a mechanism for its eventual dissolution.

WHITE CELLS (Leucocytes)

The leucocyte count in normal blood varies between 4×10^9 and 11×10^9 cells per litre of blood. The cells are all polymorphonuclear leucocytes which have cytoplasmic granules which react with Romanowsky stains. This allow the identification of three types, collectively known as granulocytes:

Neutrophil
Eosinophil
Basophil.

Neutrophil leucocyte

Neutrophils form about 60% of circulating white cells. They exist for about 8–10 h in the circulation and pass into tissues. Once there they do not return. The nucleus is divided into a number of lobes joined by a thin chromatin strand. Granules in cytoplasm stain pale pink and the neutrophils exist as two pools:

1. A marginal pool in loose contact with the walls of blood vessels.
2. A circulating pool.

The primary function of the neutrophil is to defend the body against infection by phagocytosis of the invading organisms, which are attracted to them by a process known as chemotaxis.

Eosinophils

Eosinophils make up about 1–6% of circulating white cells. They have large cytoplasmic granules staining red and a bilobed nucleus. They circulate in the blood for about 4–5 h before entering the tissues. There they play a special role in parasitic infestations and in the allergic response.

Basophils
These are the least numerous of the white cells, about 1%. They have large cytoplasmic granules which stain blue and contain heparin and histamine. They are involved in allergic reactions of the atopic or anaphylactic type and when in contact with immunoglobulin E (IgE) release the granules stored in the cytoplasm.

Monocytes
Monocytes make up about 2–8% of white cells. A monocyte is one of the largest cells in the blood (12–20 μm). The nucleus is oval or horseshoe-shaped in a cytoplasm with pale azurphilic granules. They circulate in the blood for about 10 h before entering the tissues where they become tissue macrophages. They are actively phagocytic and remove senile red cells. They also function (with T-lymphocytes) to promote the stimulation of the immune system by antigens.

Lymphocytes
Lymphocytes make up about 20–30% of the white cells. They have a dense nucleus which almost fills the cytoplasm. They are often described as small or large lymphocytes with the latter having more cytoplasm, though they are smaller than monocytes. There are many different types which play a part in immunity:

- T-lymphocytes (40–80% of blood lymphocytes) which are responsible for cell-mediated immunity.
- B-lymphocytes (10–30%) which are responsible for humoral immunity.

The platelets
Platelets are cytoplasmic fragments, without a nucleus derived from the megakaryocyte. They circulate as discs, 2–4 μm in diameter, 0.6–1.2 μm in thickness and have a volume of 6–9 fl. The platelet count is $150–400 \times 10^9$/L of blood. A circumferential band of microtubules maintains its shape, but there is extensive invagination of the plasma membrane. The plasma membrane contains phospholipid (platelet factor 3) and arachidonic acid. A tubular system is present which stores Ca^{2+} which is released into cytoplasm when the platelets are activated. Actin–myosin filaments are present together with organelles, lysosomes and mitochondria. Also present are:

Glycogen granules
Dense granules containing ADP, 5-HT and Ca^{2+}
α-granules containing fibrinogen, von Willebrand factor, factor V, heparin-neutralizing factor
β-thromboglobulin, platelet-derived growth factor.

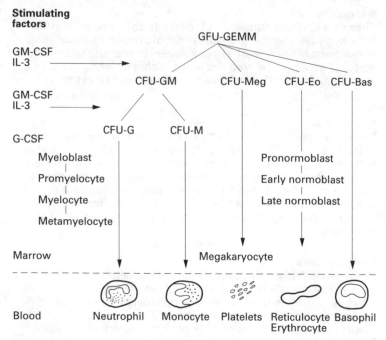

Fig. 1.2
Blood cell stimulating factors.

Granulopoiesis and monopoiesis

Granulopoiesis and monopoiesis are the processes involved in the growth and development of polymorphonuclear granulocytes and monocytes. The following abbreviations are used:

CFU-GM = colony-forming unit producing granulocyte-macrophages.
CFU-Eo = colony-forming unit producing eosinophils.
CFU-Bas = colony-forming unit producing basophils.
CFU-Meg = colony-forming unit producing megakaryocytes.

Colony-stimulating factors (CSF) (Fig. 1.2)
GM-CSF = growth factor stimulating the production of neutrophils, eosinophils, monocytes, red cells and platelets.
G-CSF = stimulates production of neutrophils.
M-CSF = stimulates the production of monocytes.
IL = interleukins numbered 1–7.

The following growth factors are required for the maturation and development of the:

Eosinophil GM-CSF and IL-5
Basophil IL-3 and IL-4
Neutrophil GM-CSF and G-CSF
Monocyte GM-CSF and M-CSF.

Thrombopoiesis

Megakaryocytes are formed from CFU-Meg by a process called endomitotic replication. In this DNA replication and an increase in cytoplasmic volume occur, but not cellular division. Megakaryocytes may have up to 32 times the normal diploid (2n) content of other cells. The cell then becomes lobulated and the cytoplasm matures with the formation of platelets which are shed into the venous sinuses. Each megakaryocyte produces between 2000 and 7000 platelets. When the number of circulating platelets falls a substance called thrombopoietin is produced which stimulates the production of megakaryocytes.

Lymphopoiesis

CFU-L, the lymphoid stem cell, matures and differentiates along two different pathways, giving rise to two main populations of lymphoid cells. These are the T-lymphocytes which constitute about 40–80% of the blood lymphocytes and the B-lymphocytes which constitute about 10–30%. A small population of cells belonging to neither group also circulates. These are called null cells. T-lymphocytes depend upon the thymus for their maturation, while B-lymphocytes differentiate in the fetal liver and adult bone marrow.

THE HEART

The heart is one of the major organs in the body. Its primary function is to pump blood into the arterial system. It has its own intrinsic rhythm, basal heart rate, which can be increased or decreased by neural and hormonal factors.

CARDIAC MUSCLE AND ITS PROPERTIES

Structure
The heart has four chambers:
Left atrium
Right ventricle
Right atrium
Left ventricle.

Major vessels
Vena cava

Pulmonary artery
Pulmonary veins
Aorta.

Blood supply
Cardiac muscle blood supply comes from coronary circulation
arising from two arterial branches in the aorta.

Valves
Valves separate the chambers of the heart, enabling unidirectional
blood flow. The valves are:
Tricuspid – lies between the right atrium and right ventricle.
Mitral – lies between the left ventricle and left atrium.
Pulmonary – lies between the right ventricle and pulmonary artery.
Aortic – lies between the left ventricle and the aorta.

Innervation
The heart is innervated by the autonomic nervous system:
Sympathetic nerves releasing noradrenaline.
Parasympathetic nerves releasing acetylcholine.

Hormonal influences
Adrenaline released from the adrenal medulla.

DIFFERENT TYPES OF MYOCARDIAL CELL

Cells in different regions of the heart perform different functions
and, consequently, have different structures, electrical and
contractile properties.

Sinoatrial (SA) node
Small specialised cells (10 μm in length) lying in the right atrium
close to the vena cava.
Cells form a discrete but loose network of interconnected cells.
Spontaneously active firing at the highest frequency of all heart
cells.
Brief action potential (100 ms), overshoot to +10 mV.
Low actin–myosin content.
Weakly contractile.

Atria
High actin–myosin content.
Strongly contractile.
Brief action potential – slow upstroke.
Action potential duration – 100 ms, overshoot +20 mV.
Little or no pacemaker activity.
Connected to SA node through intercalated discs (gap junctions).
Atrial cells are coupled and function as a syncytium.

Atrioventricular (AV) node
Specialised cells connecting the atrial muscle to the bundle of His through the fibrous ring.
Poorly contractile.
Pacemaker activity is slower than the SA node.

Bundle of His
Discrete specialised cells (Purkinje cells) running from the AV node to the apex of the ventricles.
Travel as two discrete left and right branches.
Fast conducting.
Action potentials with a rapid upstroke and long plateau.
Action potential duration 200–300 ms, overshoot +40 mV.
Pacemaker activity but slower than the SA node.
Low actin–myosin content.
Weakly contractile.
Function is to carry excitation to the apex of the heart so that it contracts first.

Ventricle
High in actin and myosin.
Strongly contractile.

Other functional features in the heart

Fibrous ring
A region of non-cellular material which separates the atria from the ventricles.
Functions to support the main valves in the heart into which the myocardium attaches.
A further function is to prevent direct spread of the atrial action potential to the ventricles.

Intercalated discs
Specialised regions connecting adjacent cardiac myocyte structures are similar to gap junctions.
Cells are separated by a gap of 2–4 nm which is bridged by the junctional proteins.
The junctions form a low resistance pathway between cells allow rapid conduction of the cardiac action potential.

STRUCTURE OF THE CARDIAC MYOCYTE

The cardiac myocyte (Fig. 1.3) has many structural similarities to skeletal muscle.
1. The cells are surrounded by a plasma membrane in which there are voltage-operated ion channels and other proteins associated with the transport of a large number of substances.
2. A transverse tubular network (T-tubules) is connected to the

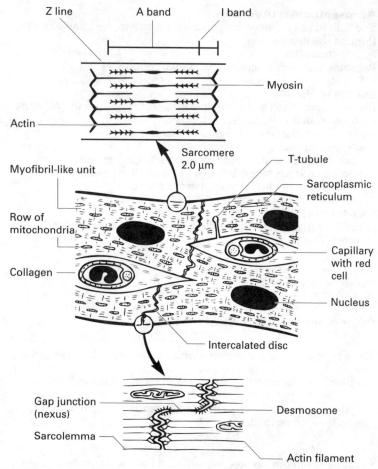

Z line   A band   I band

Myosin

Actin

Sarcomere
2.0 μm

Myofibril-like unit

T-tubule

Sarcoplasmic
reticulum

Row of
mitochondria

Collagen

Capillary
with red
cell

Nucleus

Intercalated disc

Gap junction
(nexus)

Desmosome

Sarcolemma

Actin filament

Fig. 1.3
The cardiac myocyte. Section of myocardium parallel to fibre axis, based on electron
microscopic studies. The widths of the sarcomere (2 μm) and red cell (7 μm) indicate
the scale. The enlargement of part of the intercalated disc at the bottom shows a gap
junction or nexus along the horizontal step (interplicate segment) and two
desmosomes. (Reproduced with permission from: Levick JR 1995 An introduction to
cardiovascular physiology, 2nd edn)

plasma membrane and extends into the myocyte. The T-tubules in
cardiac cells are not as well developed as those in skeletal muscle.
3. There is a second system of intracellular membranes –
   sarcoplasmic reticulum (SR) – which is not directly linked to the
   T-tubules. As in skeletal muscle, the T-tubules and SR are in

close apposition to each other. The contact between the T-tubules and SR is not the same as in skeletal muscle and plays a different role in excitation-contraction coupling.
4. Lying between the SR are the myofibrils of the contractile proteins – actin and myosin. The actin filaments are also associated with tropomyosin and troponin as in skeletal muscle.
5. The myocytes contain large numbers of mitochondria.

THE CARDIAC ACTION POTENTIAL

The action potential is the electrical signal which travels throughout the cardiac muscle to initiate contraction. The membrane potential of cardiac cells is never stable. In general the action potential involves:
- A rapid depolarisation of the membrane when the potential approaches –50 mV – the rising phase.
- The peak of the action potential is close to +40 mV.
- After the peak the membrane potential partially repolarises to around +10 mV for a prolonged period (200 ms) – the plateau phase.
- The potential then rapidly returns to a negative value near –70 mV.
- There is then a slow depolarisation – the pacemaker potential which takes the membrane potential towards threshold.
- The cycle repeats.
Although the form of the action potential is similar in all cardiac cells, there are distinct differences.

IONIC BASIS OF THE CARDIAC ACTION POTENTIAL

The cardiac action potential is in many ways similar to the nerve impulse, but lasts for much longer. The duration can be as long as 300 ms. The different phases of the cardiac action potential and the mechanisms underlying them are:
- The rising phase of the action potential is due to a rapid influx of Na^+ and Ca^{2+} resulting from the activation of voltage-operated ion channels (VOC).
- The Na^+ channels close rapidly but the Ca^{2+} channels remain open for some time.
- This Ca^{2+} influx keeps the membrane depolarised and in the plateau phase.
- It is during this phase that the Ca^{2+} entry triggers contraction.
- Towards the end of the plateau phase the Ca^{2+} channels begin to close and K^+ channels begin to open.
- Opening of K^+ channels leads to rapid repolarisation and the membrane potential returns to a negative potential.
- The K^+ channels begin to close resulting in a gradual depolarisation – pacemaker potential.
- The pacemaker is augmented by a gradual increase in Na^+ and Ca^{2+} influx.

- At threshold the VOC for Na^+ and Ca^{2+} channels undergo regenerative opening.
- The cycle repeats.

INITIATION OF THE HEART BEAT

The cardiac cycle involves a repeated series of events. This coordinated activity is essential for the proper working of the heart as a pump. The sequence of events is:
- The SA node fires action potentials at the highest intrinsic frequency and therefore drives all of the other cardiac cells.
- The frequency of firing depends on the pacemaker potential – a faster pacemaker depolarisation takes the membrane potential to threshold faster and therefore initiates an action potential sooner. This is the process where chronotropic agents, acetylcholine and noradrenaline, have their action.
- The atrial cells are activated by the SA node cells. The SA node action potential is conducted to the atrial cells using the intercalated discs, from where it spreads to every atrial cell to initiate atrial contraction.
- The ring of fibrous material which divides the atrium from the ventricles does not conduct action potentials and, consequently, the atrial action potential does not directly activate the ventricles.
- The AV node is activated by atrial electrical activity which results in activation of the Purkinje cells. Conduction is rapid in the Purkinje cells and the action potential is rapidly transmitted to the ventricular myocytes in the apex of the heart.
- The apex ventricular cells are excited and the action potential spreads upwards towards the fibrous ring.
- The electrical events in each cell are used to initiate contraction.

EXCITATION-CONTRACTION COUPLING (Fig. 1.4)

Excitation-contraction coupling involves the cardiac action potential and the subsequent events, resulting in contraction of the cardiac muscle cells. The sequence of events is:
- During the plateau phase of the action potential Ca^{2+} enters the cytoplasm via the VOC Ca^{2+} channels.
- This initial rise in Ca^{2+} acts on specific proteins on the surface of the terminal regions of the SR (ryanodine receptors). These receptors are part of Ca^{2+} channels which open to allow Ca^{2+} to move from the SR into the cytoplasm.
- This process is called Ca^{2+}-induced Ca^{2+} release.
- Ca^{2+} diffuses to the troponin C protein to reveal the myosin binding site.
- A cycle of cross-bridge attachments, ATP binding and hydrolysis takes place to facilitate the production of force.

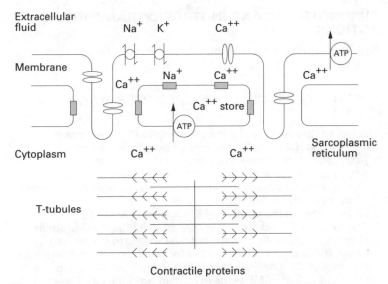

Fig. 1.4
Excitation-contraction coupling in cardiac myocytes. As the action potential begins the plateau phase, Ca^{2+} channels are activated which allow Ca^{2+} ions to enter the cytoplasm. This Ca^{2+} influx triggers the release of Ca^{2+} stored in the sarcoplasmic reticulum by acting on Ca^{2+} release sites. The total rise in Ca^{2+} is then responsible for activation of the contractile apparatus. At the end of the action potential Ca^{2+} in the cytoplasm is taken back into the sarcoplasmic reticulum by ATP-driven pumps or extruded from the cell by ATP-dependent pumps on the surface membrane or the Na^+/Ca^{2+} exchange mechanism.

- Ca^{2+} is removed from the cytoplasm by ATP-dependent Ca^{2+} pumps on the SR, Ca^{2+} pumps on the plasma membrane and in exchange for Na^+ ions in the external medium (Na^+/Ca^{2+} exchange).
- As the cytoplasmic Ca^{2+} concentration falls the contractile proteins are not activated and force production stops.
- The muscle relaxes ready for the next cycle.
- The cardiac action potential is very long (200–300 ms) and contraction is initiated by Ca^{2+} influx during the plateau phase. As the action potential repolarises the Ca^{2+} influx stops and contraction is terminated. As with the nerve and skeletal muscle action potential, the cardiac myocyte has a refractory period when no further action potentials can be initiated and consequently no contraction. The long action potential and the refractory period ensure that the heart contracts and relaxes. This is essential to allow the chambers to fill during relaxation and empty during contraction.

CHRONOTROPIC AND INOTROPIC TRANSMITTER ACTIONS

Both sympathetic and parasympathetic nerves innervate cardiac muscle.

ACETYLCHOLINE-MUSCARINIC RECEPTOR ACTIVATION
(Fig. 1.5)

The transmitter released by the postganglionic parasympathetic nerves is acetylcholine. The actions of acetylcholine include:
- Activation of muscarinic receptors on the myocytes.
- Act on the SA node cells to slow the pacemaker potential.
  The pacemaker potential is affected by two distinct mechanisms:
1. A decrease in the influx of Na^+.
2. An increase in the number of K^+ channels operating.

The muscarinic receptors are coupled to inhibitory G-proteins which are linked to the enzyme adenylate cyclase decreasing its activity. This enzyme is responsible for producing cyclic adenosine monophosphate (cAMP). cAMP is needed to open Na^+ channels, therefore a lack of cAMP leads to a fall in Na^+ influx and less depolarisation. The pacemaker potential is slowed.

The muscarinic receptors are also linked to K^+ channels via a G-protein. Receptor binding leads to an opening of K^+ channels and a hyperpolarisation of the membrane potential. This shift in potential reduces the likelihood of the potential reaching threshold and initiating an action potential. The pacemaker potential is slowed.

NOR-ADRENALINE-β_1 ADRENORECEPTOR ACTIVATION

Activation of the sympathetic nerves leads to the release of nor-adrenaline and a consequent increase in heart rate (chronotropic action) and increase in the force of contraction of each heart beat (inotropic action).

The chronotropic effect is brought about by an increase in the rate of depolarisation of the pacemaker such that threshold for the initiation of an action potential is reached more rapidly. These changes are the result of activation of an increased influx of Na^+ and Ca^{2+} ions. Adrenoreceptor activation also reduces the conduction delay at the atrio-ventricular node and accelerates the sequestration of Ca^{2+} by the sarcoplasmic reticulum.

The inotropic actions are mediated by an increase in the magnitude of the Ca^{2+} current during the action potential.

Activation of β_1 adrenoreceptors leads to the activation of G proteins and the stimulation of adenylcyclase. The resulting rise in cAMP and activation of protein kinase phosphorylates the ion channels carrying the Na^+ current responsible for the pacemaker and the Ca^{2+} current triggering contraction increasing the amplitude of both (see Fig. 1.5).

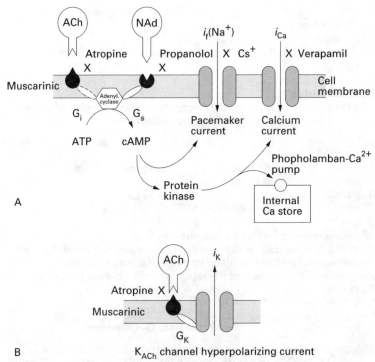

Fig. 1.5
Mechanisms by which the autonomic neurotransmitters noradrenaline (sympathetic) and acetylcholine (parasympathetic) alter the rate and force of the heart beat. A, Effect on pacemaker current (i_f) and on calcium current (i_{Ca}). G_s = stimulatory guanosine triphosphate-binding protein; G_i = inhibitory GTP-binding protein. B, Effect of acetylcholine on potassium current. (Reproduced with permission from: Levick JR 1995 An introduction to cardiovascular physiology, 2nd edn)

THE CARDIAC OUTPUT

The cardiac output is the volume ejected by a ventricle in 1 min. Cardiac output depends on heart rate and stroke volume. The stroke volume is the output of a ventricle/beat.

Cardiac output = stroke volume × heart rate

Output varies with:

Age
Sex
Activity
  lying or standing
  exercise
  sleep

Meals
Excitement and fear
Pregnancy
Disease, e.g high in hyperthyroidism and low in myxoedema
Body size.

Cardiac index
The cardiac index is the cardiac output per square metre of body surface (average is 3 L per min per square metre).

Normal values

	At rest	**Exercise**
Man	4–7 L/min	20 L/min
Top athlete		30–35 L/min
Woman	about 20% less	

Measurement of cardiac output
The measurement of cardiac output is an important physiological measurement and can be used to diagnose many clinical conditions. One method of measuring cardiac output is the Fick Principle:

$$\text{Cardiac output} = \frac{O_2 \text{ consumption}}{\text{arteriovenous } O_2 \text{ difference}} \times 100$$

O_2 consumption (measured by spirometry) = 250 ml/min
Arterial blood O_2 content (measured by arterial puncture) = 19 ml/dl
Mixed venous blood O_2 content (measured by cardiac catheter) = 14 ml/dl
Cardiac output (CO) = 250/(19–14) $\times$ 100 = 5000 ml

THE HEART AS A PUMP
ATRIAL FUNCTION

The atria act as thin-walled reservoirs and receive blood from:
• The venae cavae.
• The coronary sinus (from the veins of the heart).
   The left atrium receives blood from the pulmonary veins. Atrial systole begins at the peak of the P wave of the ECG, with the right atrium contracting slightly ahead of the left atrium. Regurgitation from the right ventricle into the right atrium is prevented by the tricuspid valve, which is restrained from being everted by the chordae tendinae attached to the ventricular wall by the papillary muscles. Blood leaves the atria in two phases:
1. At slow heart rates: a large rapid passive phase and a small active phase due to atrial contraction (atrial boost).
2. At fast heart rates: an active atrial contraction becomes

important which is responsible for 10–40% of the increase in cardiac output.

Atrial contraction is responsible for the 'A' wave of the jugular venous pulse. The reverse movement of blood from left ventricle into the left atrium is prevented by the mitral valve (bicuspid), which is also retained by chordae tendinae and papillary muscles.

VENTRICULAR FUNCTION

The duration of filling, in a resting man (about 0.5 s), has two phases corresponding to the ejection phases of the atria.
Contraction begins after the QRS complex on the ECG is completed and is divided into two phases:

Isometric contraction

Wall tension rises with a rapid increase in intraventricular pressure. The atrioventricular (AV) valves close and are held in apposition by chordae tendinae and papillary muscles, and so for a short time (0.05 s) the contraction is isovolumetric as the semilunar valves are also closed.

Ejection phase
Duration about 0.3 s.
When ventricular pressure exceeds aortic and pulmonary pressures, the semilunar valves (aortic and pulmonary) open and blood is ejected into the aorta and pulmonary artery.
Rapid ejection phase (two-thirds of the stroke volume).
Slower ejection phase.
Ventricles relax.
Semilunar valves close as pressure in aorta exceeds that of the ventricles.

Isometric relaxation

For a while (about 0.08 s) the ventricles relax, tension in the walls falls and ventricular pressure falls, but with the semilunar valves and AV valves closed (isovolumetric relaxation).
AV valves open as ventricular pressures fall below atrial pressures and blood, which had been accumulating in the atria during ventricular contraction, begins to enter ventricles.

Figure 1.6 shows the pressures in the heart chambers during the cardiac cycle.

ASSESSMENT OF THE FUNCTION OF THE HUMAN HEART

CLINICAL EXAMINATION

Physical examination is carried out by routine methods of inspection, palpation, percussion and auscultation.

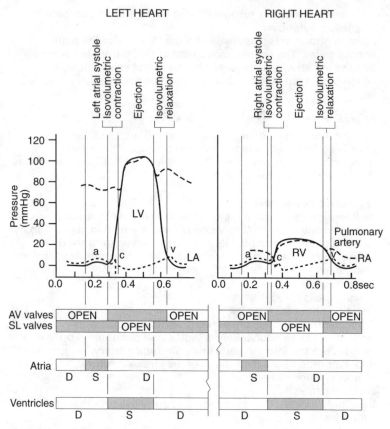

Fig. 1.6
The pressures in the heart chambers during the cardiac cycle on the left and right sides of the heart in man. AV = atrioventricular; D = diastole; LA = left atrium; LV = left ventricle; RA = right atrium; RV = right ventricle; SL = semilunar.

Pulsations on the chest wall

Apex beat of the heart

The apex beat of the heart is found within the midclavicular line in the 5th intercostal space. A complex rotary movement of the heart as it contracts produces the apex beat. Its position, force and quality will tell the physician information about the heart. The heart moves with each phase of the cardiac cycle:

Diastole – soft and pliable.
Systole – hard with an increase in anteroposterior and a decrease in transverse diameter.

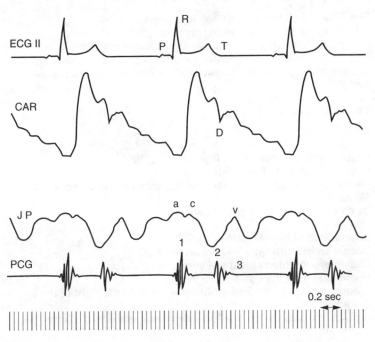

Fig. 1.7
Simultaneous records from a normal male aged 22 years. ECG = electrocardiogram, standard lead II; CAR = external carotid artery; JP = jugular phlebogram; PCG = apical phonocardiogram. D marks the dicrotic notch; a, c and v mark the three main peaks of the phlebogram; 1, 2, 3 indicate the first, second and third heart sounds respectively.

Pulsations in the neck
The movement of blood in the major arteries in the neck can be detected by touch. Two pulses may be observed at the root of the neck: the arterial and venous (jugular pulse).

Arterial pulse
A single pulsation seen with each beat of the heart sharp in onset which lifts the finger when palpated.

Venous or jugular pulse
This has more than one peak and does not lift the finger. It is best seen when the subject is supine and relaxed with the head on a pillow. Three waves may be seen but only two are prominent. They have been named:

The *'A'-wave* (A for atrial) is caused by contraction of the right atrium and the reflux of blood through the superior caval opening to slightly raise the venous pressure.

The *'C'-wave* (C for carotid) has two causes:
  bulging back of the tricuspid on closing;
  transmission from the nearby carotid artery in the neck.
The *'V'-wave* (V for ventricular systole) is caused by the rapid rise in atrial pressure when the tricuspid is shut during ventricular systole.

The heart sounds

Unaided ear
1st sound:
Loudest and longest and of low pitch – 'LUB'
Occurs at the end of the QRS complex
Causation – abrupt rise in ventricular pressure and closure of the AV valves
Best heard nearest the apex beat.
2nd sound:
Shorter and of higher pitch – 'DUPP'
Occurs shortly after the T-wave on the ECG
Causation – sudden closure of semilunar valves
Heard best near the base of the heart:
  aortic valve – in the 2nd intercostal space to the right of the sternum;
  pulmonary valve – in the 2nd intercostal space to the left of the sternum.
The 2nd sound is often split which widens during inspiration – 'DUPP' becomes 'TRUPP'.

Aided ear
With a stethoscope.

3rd sound:
Heard in children and young adults
Occurs at the end of the rapid phase of ventricular filling
Causation – rush of blood into the relaxing ventricles.

4th sound:
Inaudible in normal subjects
Present on the phonocardiogram (PCG)
Occurs in the PR interval of the ECG
Causation – contraction of atria and flow through the AV valves.

MYOCARDIAL METABOLISM

Cardiac metabolism is mainly aerobic. The immediate source of energy for contraction is ATP, synthesised by oxidative phosphorylation in the mitochondria. The reserve of high energy bonds is in the form of creatinine phosphate. O_2 is extracted from the coronary blood to an extent of 65–75% at rest; during heavy exercise this may rise to 90%.

OXYGEN UTILISATION BY THE HEART

	Extraction rate	Arterial blood content (ml/L)	Coronary sinus blood content (ml/L)
At rest	65–75%	195	50–70
Heavy exercise	90%	195	20

Further demand for oxygen is met by an increase in coronary blood flow. For a given work load, O_2 consumption is greater when the myocardium is working against a high blood pressure (pressure load) than an increased diastolic volume (volume load).

METABOLIC SUBSTRATES

Free fatty acids supply 65–70% of energy requirements, and more during endurance exercise:
Glucose – 11 g/day
Lactose – 10 g/day.
The heart increases its utilisation of whatever substrate is most abundant, e.g.: in uncontrolled diabetes, ketone bodies; in hard exercise, lactic acid. Oxidation of lactate involves using the enzyme lactic dehydrogenase of a type specific to the heart (LDH4), so that after an infarction (death of cardiac muscle in a heart attack) this enzyme, together with creatine phosphokinase and aspartate aminotransferase, escape into the circulation and provide a diagnostic test to prove the heart muscle is damaged.

STARLING'S LAW OF THE HEART

'The energy of contraction of a cardiac muscle fibre, like that of a skeletal muscle fibre, is proportional to the initial fibre length, at rest.' In other words, the greater the stretch of the ventricle in diastole, the greater the stroke work achieved in systole. Starling demonstrated:
1. If diastolic filling was increased, the heart increased its stroke volume.
2. A rise in arterial resistance (pressure) increased the work carried out by the ventricle at an increased diastolic volume.

VENTRICULAR PERFORMANCE

Ventricular performance is often described using *ventricular performance curves* (Fig. 1.8). Ventricular performance curves have for their ordinate either stroke volume or stroke work (stroke volume × mean arterial pressure), and for their abscissa, some index of the cardiac fibre length, e.g. ventricular end diastolic pressure (EDP) or end diastolic volume (EDV).

Pressures may be measured by direct cardiac catheterisation but volumes are measured indirectly, by some angiographic technique

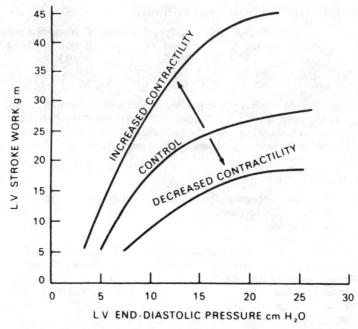

Fig. 1.8
Ventricular performance curve. (Reproduced from: Braunwald E, Ross J, Sonnenblick EH 1967 Mechanisms of contractions of the normal and failing heart. J & A Churchill, London.)

or echocardiography. As the right ventricular end diastolic pressure is nearly equal to the central venous pressure (CVP), the latter is used in the human as it is relatively easily measured by cardiac catheterisation.

VENTRICULAR FILLING

Ventricular filling is determined by the venous return to the heart which also sets the central venous pressure which is approximately equal to the end diastolic pressure and determines the end diastolic volume.

FACTORS INFLUENCING THE CENTRAL VENOUS PRESSURE (CVP) AND THE VENOUS RETURN TO THE HEART

Blood volume

The greater the blood volume the greater the average venous

pressure, this comes about because about two-thirds of the blood volume is located in the veins.

The action of the heart as a pump
The heart transfers blood from the venous system into the arterial system, thus raising pressure in the arteries and causing the CVP to fall if it is not corrected.

The respiratory pump
Intrathoracic pressure is negative (-5 cmH$_2$O) and, during inspiration, becomes more negative (-10 cmH$_2$O). The descent of the diaphragm also increases the intra-abdominal pressure, both act by promoting the filling of the central veins by increasing the venous pressure gradient.

Gravity
On standing there is a pooling of blood into the veins of the legs, with about 500 ml of blood being redistributed so the CVP falls and, consequently, the stroke volume falls. An increase in end diastolic volume brought about by an increase in CVP will increase cardiac output. A fall in end diastolic volume causes a drop in cardiac output.

Peripheral venous tone
The amount of blood in the peripheral veins (skin, splanchnic area and kidneys) is in part determined by the degree of venoconstriction which in its turn is controlled by the sympathetic nerves which innervate them. An increase in venomotor tone decreases their capacity and increases the CVP (i.e. like an increase in blood volume); venodilatation has the reverse effect.

The muscle pump
During movements and especially during exercise the muscles squeeze the veins and displace blood centrally. Back flow in the superficial veins is prevented by their valves.

Effect of ventricular filling on the stroke volume
The end diastolic volume is determined by all those factors above which influence the return of blood to the heart and upon the distensibility of the ventricle. As the CVP rises so the stroke volume increases. The Starling mechanism ensures that the output from the two sides of the heart is equal over a period of time. An increase in CVP increases the output on the right side of the heart. Pressure in the pulmonary circulation is transmitted through the pulmonary veins to the left ventricle to increase left ventricular end diastolic pressure and the left ventricular stroke volume. Thus differences in stroke volumes between right and left ventricles are only transient, lasting for a few beats only, i.e. left and right ventricular outputs are balanced.

Effects of arterial pressure on the stroke volume

The effect of a rise in arterial pressure depends upon the interplay between a number of factors. A rise in arterial pressure caused by vasoconstriction will oppose the emptying of blood into the aorta and the stroke volume will fall. This is for only a few beats of the heart for, as the arterial pressure rises, the stroke volume initially falls for a few beats but:

- The venous return continues.
- The diastolic volume will therefore increase.
- Energy for contraction is increased by the Starling mechanism.
- The output is restored within a few further beats.

In the intact animal or person there is a further factor to consider. When the arterial pressure rises the baroreflex is invoked (see p. 55) which reflexly reduces activity in the sympathetic nerves to the heart. This reduces the heart rate and contractility of the ventricles.

The contractile force of the ventricles is controlled by two mechanisms:

1. *Intrinsic regulation* (within the myocyte) – the Starling mechanism.
2. *Extrinsic regulation* or *inotropic state* (increased contractility), brought about by agents outside the myocyte, such as:
   a. noradrenaline;
   b. adrenaline;
   c. activation of the sympathetic nerves to the heart.

An increase in sympathetic nerve activity (e.g. exercise, orthostasis, stress, haemorrhage) increases the contractile energy produced at any given diastolic length of the ventricular myocyte (see Fig. 1.8). Angiotensin II, which acts by facilitating the release of noradrenaline from sympathetic nerves and by enhancing the Ca^{2+} plateau current (see p. 21), augments the rise in intracellular Ca^{2+}. Inotropic agents cause a more forceful and faster contraction, i.e systole is shortened. Ventricular pressure rises more rapidly, to an increased peak pressure. The ejection fraction is increased. There is a transient increase in stroke volume which is limited as there is a concomitant fall in end diastolic pressure and a rise in arterial pressure. If the venous return increases as in exercise, then the inotropic effect causes a large increase in stroke volume.

THE BLOOD VESSELS

CLASSIFICATION OF VESSELS

Functional	Anatomical
Damping	Arteries
Resistance	Small arteries
	Arterioles
Exchange	Capillaries
Capacity	Veins and venules

STRUCTURE

All vessels have the same basic structure with the exception of the exchange vessels. In cross-section they can be seen to consist of three coats:

Tunica intima
Tunica media
Tunica adventitia.
   These three coats vary in their composition, depending upon the nature of the function of the vessel.

ARTERIAL OR DAMPING VESSELS

These are elastic arteries (diameters 1–2 cm) and are the aorta, the pulmonary arteries and their major branches. They have the following basic structure:

Tunica intima
A flat endothelial layer made up of tapering squamous cells whose long axis lies in the direction of the vessel.
An internal elastic lamina with collagen fibrils.

Tunica media
Concentrically arranged elastic laminae.
Smooth muscle cells between adjacent laminae.
A meshwork of collagen fibres.

Tunica adventitia
Mainly collagen fibres.

Function
The walls are very distensible due to the elastic laminae which dampen pressure fluctuations caused by the beating of the heart. In systole the elastic tissue is stretched and energy is stored as potential energy. In diastole, there is an elastic recoil which limits fall in pressure between each beat and an intermittent ejection of blood is converted into a continuous flow. The role of smooth muscle is protective, i.e. it prevents overdistension and contracts on damage preventing excessive blood loss.

RESISTANCE VESSELS

Resistance vessels can be divided into two classes:

Small muscular arteries (0.1–1 cm)
These are small arteries, such as the radial artery, which act as low resistance conduits and are characterised with a tunica media

which is thick (relative to the lumen diameter) and muscular. About one-half of the precapillary resistance lies in these vessels proximal to the arterioles.

Tunica intima
Endothelial cells as in other blood vessels with elastic fibres and collagen fibrils.

Tunica media
Bounded on the luminal side by a well-defined internal elastic lamina. The external lamina is either fragmented or missing. Smooth muscle layers, circumferentially arranged, decreasing in number as the diameter of the vessel decreases. They appear to form an electrical syncytium.

Tunica adventitia
Contains connective tissue (elastin and collagen), fibroblasts, mast cells.
Adrenergic unmyelinated axons with interspersed varicosities at intervals and surrounded by Schwann cells. They do not penetrate into the media.

Arterioles (small arteries <30–50 μm)
These are similar in structure to the small arteries. By dilating and constricting their major role is to control the blood flow to the capillaries according to local need. They are the site for the greatest control of the peripheral resistance.
Peripheral resistance varies as the diameter of the arteriole.

Poiseuille's Law
The rate of flow through a rigid tube: $= \dfrac{P_1 - P_2}{8L} \times \dfrac{r^4 \pi}{\eta}$

where:
$P_1 - P_2$ = pressure difference between ends
L = length of tube
r = radius of tube
η = viscosity of blood
π = 3.142

$$\text{Resistance to flow} = \frac{8\eta L}{\pi r^4}$$

Note that the radius of a tube is the most important factor in determining the resistance to flow, e.g. a two-fold decrease in radius increases the resistance to flow 16 times ($2^4 = 16$).

Factors affecting vessel calibre
There are several factors which all interact to contribute to the control of vessel diameter:

Intrinsic control (within the vessel wall)
Myogenic autoregulation – in some organs large changes in
perfusion pressure cause little or no change in blood flow.

Metabolic control
Relaxation of arterial smooth muscle in response to:

Local decrease in Po_2.
Local increase in Pco_2.
Local increase in $[H^+]$.
Increase in temperature.
Local increase in many metabolites, e.g. ATP, ADP, AMP, phosphate,
lactate, pyruvate.
$[K^+]$: increased osmolality.

Humoral (chemical or autocoid) control
Endothelium-derived relaxation factor (EDRF)→vasodilatation.
Kallikrein→kinins (e.g. bradykinin)→vasodilatation in glands and
skin (see under salivary gland, p. 115).
Histamine (released from mast cells)→vasodilatation in
inflammatory reactions.
Prostaglandins (PGFs→vasoconstriction; PGEs→vasodilatation).
Leukotrienes (in inflamatory responses)→vasoconstriction.
Thromboxane A_2 and 5-hydroxytryptamine released in the clotting
mechanism→vasoconstriction.

Hormonal control
Catecholamines – adrenaline (activates both α- and
β-adrenoreceptors) the response depends upon the relative
densities of the two receptors in the vessels of the tissue under
consideration.
Angiotensin II→vasoconstriction (see p. 297).
Vasopressin→vasoconstriction.
Atrionaturetic peptide→decreases sensitivity to vasoconstrictors.

Neural control
Largely though the sympathetic nervous system controlled by the
cardiovascular centres in the brain stem.
Neurotransmitter – noradrenaline acts on. α-adrenoreceptors→
vasoconstriction.
Parasympathetic nervous system→vasodilatation but only in
certain tissues:

  salivary glands, stomach, pancreas, colon
  external genitalia (erectile tissue) – pelvic parasympathetics.
Neurotransmitters:

  acetylcholine
  vasoactive intestinal polypeptide.

CAPILLARIES AND EXCHANGE VESSELS

These are the smallest vessels, about 4–7 μm in diameter and about 250–750 μm long, and they are the site at which major exchange between blood and the tissues takes place. However, the respiratory gases O_2 and CO_2 can exchange across the walls of the smaller arterioles and the postcapillary venules.

Factors affecting capillary pressure
Capillary pressure is increased:
Below the level of the heart
With increased venous pressure
With skin temperature.
    Capillary pressure is reduced:
On contraction of precapillary resistance vessels (precapillary sphincters).

Blood flow in capillaries
The pattern of blood flow in capillaries has the following characteristics:
* Velocity $\cong 0.5$ mm/s.
* Red cells move in jerks.
* Pulsatile flow is lost, except during intense vasodilatation.
* Red cells flow in single file in narrow capillaries.
* Red cells can temporarily undergo deformation to pass through narrow capillaries.

The microcirculation
The microcirculation consists of:
The smaller arteries
Arterioles
Metarterioles (terminal arterioles just proximal to the capillary)
Thoroughfare (preferential) channels
Capillaries
Venules
Arteriovenous anastomoses.

Arteriovenous anastomoses
These are shunt vessels which pass directly from arterioles to venules and bypass the capillary bed. They are about 20–135 μm in diameter and have thick muscular walls innervated by the sympathetic system. They are found in the skin, especially in the fingers, toes, ears, etc., and are involved in temperature regulation.

VEINS – CAPACITANCE VESSELS

Structure
They have a similar basic structure to arteries but:

Have thinner walls.
Have less elastic tissue.
Are more easily distended.
Valves are present in some superficial vessels.
Venules and small veins are more numerous than arteries and arterioles.

Function

Transport of blood at high velocity and low energy cost from the tissues back to the heart
Adjustment of the capacity of the circulation – their capacitance function is to act as a blood reservoir.

CENTRAL VENOUS PRESSURE (CVP)

The CVP represents the right heart filling pressure. It is used as an assessment of effective circulating blood volume during fluid replacement for hypovolaemic shock, septic, anaphylactic or cardiogenic shock.

Measurement

CVP is measured by inserting a catheter (under strict aseptic techniques) into a peripheral vein so that its tip lies in a great vein close to the right atrium and connected to a manometer.

Normal pressure: 0–0.55 kPa (0–4 mmHg or 0–6 cmH$_2$O) at right atrial level. For a patient in the supine position the level of the right atrium is at the intersection of a line drawn horizontally around the chest at the sternal angle and the midaxillary line.
CVP varies with:

Cardiac cycle
Respiration
Position of the patient.
CVP is decreased during:

Shock (see above)
Negative pressure breathing.
CVP is increased during:

Heart failure
Expansion of the blood volume (overtransfusing)
Straining (Valsalva manoeuvre)
Positive pressure breathing.

Jugular vein as a manometer (Fig. 1.9)

CVP can be measured at the beside as the height above the horizontal plane, through the sternal angle to which the external jugular vein is distended in a patient reclining with the head supported by a pillar. The CVP equals the pressure at the point of collapse (zero) plus the pressure exerted by the vertical column of

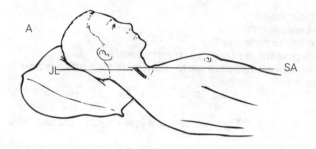

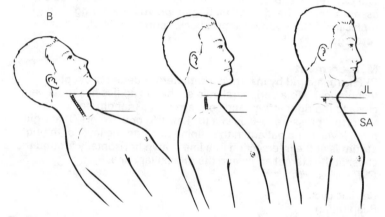

Fig. 1.9
The jugular vein as a manometer. A, The level of blood in the internal jugular vein when the subject is lying with the head supported on pillows. The upper level of the column of blood in the jugular vein (JL) lies a little below the level of the sternal angle (SA). B, The upper level of the column of blood in the internal jugular vein is above the sternal angle, indicating an increase in venous pressure. Note that as the position of the patient alters the upper level of the column of blood occupies different positions in the neck, although its relationship to the sternal angle remains the same. (Reproduced with permission from: Lewis T 1949 Diseases of the heart. Macmillan, London.)

blood between this and the right atrium. As the sternal angle is about 5 cm above the midpoint of the right atrium, then to get the CVP add 5 cm to the height of the point of collapse.

VEINS AS TRANSPORT VESSELS

Veins carry blood back to the heart. The velocity of blood increases

as it passes through venules and veins. Blood returns to the heart with very little expenditure of energy. Venous return is influenced by gravity, which:
- Aids return from above the heart – veins collapsed to a dumbbell shape and flow is confined to marginal channels.
- Impairs it below the heart.

The effects of gravity are:
- Counteracted by muscle pump.
- In superficial veins, return to the periphery is prevented by valves.

Effect of internal pressure on the volume of blood in the veins

At 1 mmHg the vein is almost collapsed with a narrow elliptical profile. From 1 to 10 mmHg the profile gradually changes from elliptical to a more rounded shape. At 10–15 mmHg the cross-section is fully circular. Venous return is influenced by the respiratory pump. Because the veins are thin walled, the fall in intrathoracic pressure on inspiration expands the veins. At the same time the diaphragm descends and compresses the intra-abdominal veins, increasing flow into the chest and heart. Conversely, on expiration these events are reversed.

Properties of veins

Veins are easily distensible. They exhibit a resting tone (a degree of active tension in the smooth muscle of the tunica media). They are also capable of contraction. This is induced by activation of sympathetic vasoconstrictor nerves. This has the effect of reducing the capacity of the veins and displacing blood into the thorax towards the heart, thus regulating the filling of the heart.

ARTERIAL PULSE

The arterial pulse is determined by:
Intermittent flow of blood from the heart.
Resistance to outflow by resistance vessels (arterioles).
Elasticity of the vessel wall.
Pulse wave velocity.

Observations made by a clinician on the pulse

Simple observations on the pulse can help the clinician with diagnosis. The following are routinely considered:

Rate
Rhythm – regular or irregular
Character
Volume ($\cong$ blood pressure)
Condition of blood vessel walls (elastic or stiff).

ARTERIAL BLOOD PRESSURE

As the heart contracts, its output is ejected into the aorta, arteries and arterioles which creates a pressure within these vessels called the blood pressure. Figure 1.10 shows the pressure measured directly from an artery through a catheter and a manometer. The blood pressure has two components, as can be seen in Figure 1.10. *Pressures in the blood vessel are defined as*:

- The *systolic pressure*, which is the maximum pressure recorded during systole.
- The *diastolic pressure*, which is the minimum pressure recorded during diastole.
- The *pulse pressure*, which is the systolic pressure minus the diastolic pressure.
- The *mean arterial pressure* = diastolic pressure + one-third of the pulse pressure.

Clinical measurement of the arterial blood pressure

For most clinical purposes the blood pressure must be measured by an indirect method called sphygmomanometry. An inflatable rubber cuff within a non-distensible cotton sleeve is wrapped snugly around the arm with the rubber bag over the brachial artery on the median side of the arm. The cuff is inflated until the pulse at the wrist disappears. With the bell of the stethoscope over the brachial artery in the antecubital fossa, the pressure in the bag is lowered slowly by allowing air to escape. No sounds are heard until

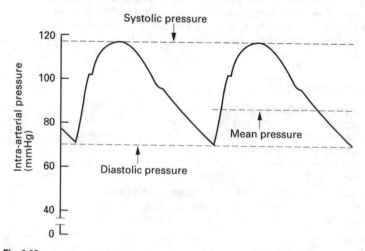

Fig. 1.10
Intra-arterial pressure in a large peripheral artery, to define systolic and diastolic pressures. Mean pressure = diastolic + 1/3 of pulse pressure.

the pressure in the bag is just below the systolic pressure when spurts of blood begin to escape through the compressed artery at the height of the systolic pressure. These are the Korotkoff sounds and are caused by turbulence created by the escaping blood. As the first tapping sound is heard, the pressure in the bag, as indicated by the level of the mercury at this point, is taken as the systolic pressure. Four phases are described. Phase I is the appearance of tapping sounds, the systolic pressure. As the pressure falls over the next 15 mmHg the tapping sounds get louder, but over the next 20 mmHg or so the sounds become quieter with a murmurish quality, this is Phase II. Over the next few (5) mmHg the sounds become louder and thumping (Phase III), and as the pressure falls further (5 mmHg) the sounds suddenly become muffled, this is Phase IV and is usually taken as diastolic pressure. After the fall of about another 5 mmHg the sounds disappear, this is Phase V. Diastolic pressure may be reported as Phase IV or Phase V.

Factors determining the arterial blood pressure
Mean blood pressure is determined by the cardiac output and the peripheral resistance.

$$BP = CO \times PR$$

Systolic pressure is affected mainly by stroke volume. Diastolic pressure is affected mainly by the total peripheral resistance and the time for the pressure to decline before the next heart beat, therefore it is somewhat dependent upon heart rate.
Systolic pressure increases when there is an:
Increase in stroke volume
Increase in ejection velocity (without an increase in stroke volume)
Increase in diastolic pressure of the previous pulse
Increase in rigidity of the arteries (arteriosclerosis).
Diastolic pressure increases when there is an:
Increase in total peripheral resistance
Increase in aortic and arterial compliance (distensibility)
Increase in heart rate.

REGIONAL CIRCULATIONS
CORONARY CIRCULATION

The myocardium is supplied by the right and left coronary arteries. They arise from the coronary sinuses above the aortic valve. Generally the left coronary artery supplies the left ventricle and anterior half of the ventricular septum, the right coronary artery supplies the right ventricle and posterior half of the ventricular septum. There is some variability with:
20% having a left coronary artery preponderance

60% a right coronary artery preponderance
30% being equally balanced.

Coronary blood flow

At rest coronary blood flow is 200 ml/min. This is 4–5% of cardiac output (60–80 ml/min/100 g).

At maximal cardiac work flow rises to 1–1.5 L/min (300–400 ml/min/100 g).

O_2 consumption is 7–9 ml/min/100 g (note this is about 20 times as great as that of skeletal muscle). This very high demand is met by a very high capillary density, which increases the surface area over which diffusion takes place and also reduces the distance to the myocardial cells. O_2 transport is assisted by myoglobin in the myocardial cells. Blood flow in the subendothelial myocardium of the left ventricle may even reverse during early systole, this is offset by a denser capillary network giving a large surface area and short diffusion pathways. The deeper myocardium relies partly on anaerobic metabolism.

Extraction of O_2 from the blood

At rest
Extraction rate from coronary blood $\cong$ 65–75%, i.e. from 19.5 ml/dl to 5–7 ml/dl.
Extraction rate for whole body $\cong$ 25%, i.e. 19.5 ml/dl to 14 ml/dl.

During exercise
Extraction rate from coronary blood can reach 90%, leaving only about 2 ml/dl of O_2.

Control mechanisms

Several factors interact to control the coronary circulation.

Mechanical factors
Flow is phasic and is reduced during myocardial contraction because branches of the coronary artery within the myocardium are compressed during systole. These are end arteries so there is little anastomotic circulation.

Left coronary flow exhibits:

- A marked fall during the isometric contraction phase
- A small sustained rise in mid-systole
- A large increase in early diastole (about 80%).

Metabolic factors
There is a strong correlation between metabolism and blood flow. Coronary blood flow increases nearly linearly with myocardial O_2 consumption. Vasodilatation of coronary vessels is caused by:

- Interstitial hypoxia

- Adenosine
- Nitric oxide
- CO_2
- H^+, K^+, PO_4^{2-}.

Interstitial hypoxia acts on the smooth muscle of arterioles, causing vasodilatation. Metabolism releases adenosine from ATP, which is a powerful vasodilator. It is not yet clear which of these is the most important but adenosine release only occurs at very low Po_2. The coronary circulation is autoregulated, which protects the myocardium against underperfusion to pressures as low as 50 mmHg.

Neurohormonal control
Activation of the sympathetic causes vasoconstriction, but as this is associated with:

> tachycardia
> increased force of contraction
> increased cardiac output
> raised arterial pressure   } increased metabolism

These changes liberate more vasodilators and counteract the constrictor effect.

Adrenaline Increases blood flow by preferentially stimulating β-receptors.

Parasympathetic (vagus nerve)
Causes vasodilatation, but this is masked by reduced metabolism:

> slow heart rate
> fall in blood pressure
> reduced O_2 consumption   } reduced metabolism

Other agents
Vasopressin→vasoconstriction
Angiotensin→vasoconstriction
Prostaglandins (release induced from coronary arteries by hypoxia)→vasodilatation.

SKELETAL MUSCLE CIRCULATION

Skeletal muscle constitutes about 40–50% of the body mass and its circulation subserves two functions:
- To supply nutrients to the muscle.
- As an important contributor to the total peripheral resistance.

Nutrition
At rest muscle receives about 20% of the cardiac output, about 1200 ml/min. During maximal exercise this rises to 90% of the blood flow (about 22 000 ml/min) and 90% of O_2 consumption

(about 3150 ml/min). 75% of skeletal muscle consists of white (twitch) fibres, as in the gastrocnemius – phasically active. 15% are tonically active red (slow) fibres, as in the soleus muscle. The tonically active, slow fibres are involved in the maintenance of posture and have a higher blood flow and a greater capillary density.

The tone of the vascular bed is controlled by:
- Neural regulators – sympathetic activity.
- Metabolic regulators.

Sympathetic innervation

Sympathetic vasoconstrictors contribute to the arteriolar tone of resting muscle through α_1-adrenoreceptors. They are controlled by the central baroreceptors and therefore play an important part in the regulation of the blood pressure. Section of these nerves causes muscle blood flow to double. The muscle veins have a sparse muscle content and have a poor sympathetic innervation and so do not respond to sympathetic activity (unlike those of the splanchnic area). At the start of exercise, adrenaline through β_2-adrenoreceptors causes an initial vasodilatation, but once exercise is established metabolic factors play a major role.

Metabolic regulators causing vasodilatation

Flow increases in parallel with local metabolism and more capillaries are opened up. The factors influencing flow include:
- K^+ (released into interstitial fluid may reach 9 mM).
- Interstitial osmolality.
- Inorganic phosphate.
- Local hypoxia (potentiates above factors).
- Endothelial derived relaxing factor (EDRF).

Effect of muscle contraction

Flow can cease during the contraction phase of rhythmic exercise but helps to empty deep veins by a massaging effect. Flow can stop completely during severe isometric contractions. On resumption of flow perfusion is increased – reactive hyperaemia.

CEREBRAL CIRCULATION

The brain is about 2–3% of the body weight, receives about 15% of the resting cardiac output and consumes about 20% of the oxygen. In addition, it consumes some 25% of the glucose of blood. Basal flow to grey matter is about 100 ml/min/100 g. Failure of the cerebral circulation results in unconsciousness in a few seconds and the death of its nerve cells within 5 min. Total blood flow is relatively constant at about 50 ml/100 g/min, but within the brain there is local readjustment to meet the metabolic demand of various groups of neurones.

Mechanisms controlling cerebral blood flow

The brain protects its blood supply by controlling flow to other organs. In order to do this the perfusion pressure is maintained and cerebral blood flow (CBF) is constant. CBF is autoregulated and is unaffected by mean arterial pressures between 50 and 150 mmHg, i.e. cerebral arterioles are constricted as pressure falls and dilated as pressure rises. Thus:

- Below 50 mmHg→signs of cerebral hypoxia (mental confusion syncope)
- Above 150 mmHg→cerebral oedema and hypertensive encephalopathy.

Mechanism of autoregulation
Autoregulation is influenced by:
  Metabolic factors   A transient reduction in blood pressure causes a transient reduction in blood flow, which allows build-up of metabolites which cause vasodilatation and restore blood flow.
  Metabolic mediators include:

- Hypercapnia or hypoxia→vasodilatation (mediated by endothelial NO production).
- Hypocapnia→vasoconstriction (e.g. hyperventilation-CBF reduced).
- H^+ and adenosine→vasodilatation.

  Myogenic factors   An intrinsic property of vascular smooth muscle is that it contracts to a rise in arterial pressure and relaxes to a fall.

Extrinsic neural influences on CBF

Intracerebral arterioles are poorly innervated by sympathetic (from superior cervical ganglia) and parasympathetic fibres (from VIIth facial nerve and trigeminal ganglion), and under normal conditions have little influence on CBF.

Regional functional hyperaemia

Different forms of mental and physical activity cause local variations in local blood flow in conscious humans, e.g. movements of the hand cause an increase in blood flow in the hand areas of the motor, pre-motor and sensory cortex. Shining a light in the eye increases blood flow in the visual cortex. Causative factors are:

- Interstitial $[H^+]$
- $[Ca^{2+}]$
- $[K^+]$
- Adenosine.

These have all been implicated linked secondarily to local glucose utilisation.

THE CAPILLARIES

The capillaries are tubular vessels of just one cell thick joined end to end. Their function is to:
- Act as exchange vessels – so their walls are highly permeable but have a low selectivity
- Connect arterioles to venules to allow a circulation of the blood.

CAPILLARY SPECIALISATION

The capillaries are divided into three types according to their structure and their specific function:

1. Continuous capillaries
These function as though they have water-filled pores, 4–9 μm in diameter, between cell junctions occupying about 0.1% of the total area.
Continuous capillaries are found in skeletal muscle, the pulmonary circulation and adipose tissue.
Because of their high permeability small water-soluble molecules exchange rapidly between plasma and interstitial fluid (ISF). Most protein is retained in the capillary.

2. Fenestrated capillaries
The adjacent membranes of the cells which make up the capillary come together in places and fuse forming large areas for permeation called fenestrations. Fenestrations can be as large as 0.1 μm in diameter and are closed by a delicate membrane.
They are found in the glomerulus of kidney, the intestinal epithelium and the choroid plexus.
Their primary function is to allow high rates of water transport.

3. Discontinuous capillaries
These have large intercellular gaps, large enough to let through macromolecules and even blood cells.
They are found in the liver, bone, marrow and spleen.
Their function is to allow proteins and blood cells to enter and leave the circulation.

PLASMA–INTERSTITIAL FLUID RELATIONSHIPS

ISF is formed from plasma through capillary pores. Two processes are involved:

1. *Filtration (bulk flow)*. This provides the fluid for the ISF but is a slow process and is not an important mechanism for the supply of nutrients to the cells.
2. *Diffusion*. This is very rapid and effective over small distances of

10 µm or less and can supply nutrients to cells some 4000–5000 times faster than bulk flow.

FILTRATION

Starling's hypothesis
Because the capillaries are highly permeable, this is a mechanism to retain fluid within the circulation. Fluid, largely devoid of protein, leaves the capillary at its arterial end and a considerable part of this is re-absorbed at the venous end. There are a number of forces acting on the fluid in capillaries.

Outward directed forces
Hydrostatic pressure created by the heart (P_c).
Colloid osmotic pressure of the proteins in the ISF – (COP_i).
Negative interstitial free fluid pressure (P_i).

Inwardly directed forces
Colloid osmotic pressure of the plasma proteins (COP_p).

The concept of filtration/re-absorption does not apply to all capillaries under all circumstances. Filtration will only occur if P_c is greater than COP_p and re-absorption only if P_c at the venous end is less than all the other inward forces.

Skin capillary pressures in humans
When measured at heart level:
   Arterial end P_c = 49 cmH$_2$O (36 mmHg)
   Venous end = 34 cmH$_2$O (25 mmHg)
   COP_p = 34 cmH$_2$O (25 mmHg).
   Thus from these figures, filtration will occur over the whole capillary.
   On standing the skin capillary pressures would have been expected to have increased from a mean 43 cmH$_2$O to 161 cmH$_2$O. The actual increase was found to be 82 cmH$_2$O.
   Thus again there would be an increased filtration over the whole length of the capillary.
   Because the increase in P_c was less than the theoretical, some regulatory mechanism is in play to stop excessive filtration.

So why do the feet not swell on standing?
Compensatory adjustments occur in vascular resistance:
• Through the baroreceptor mechanism increasing sympathetic tone
• By vascular smooth muscle contracting to an increased transmural force (myogenic).
• By a local axon reflex initiated either by increased arterial or venous pressures.
   As filtration occurs along the whole length of the capillary, fluid

must be removed entirely by lymphatic drainage in this situation. In general, all the capillaries below the level of the heart will filter fluid over most of their length or a great part of it. If vasoconstriction occurs, including vasomotion, then filtration/re-absorption (as proposed by Starling) will occur for some of the time. In capillaries above the heart, filtration/re-absorption will occur.

CAPILLARIES AT SPECIAL SITES

There are specialised sites in the cardiovascular system where the capillaries are specialised:
- Lungs – hydrostatic pressure (P_c) is less than colloid osmotic pressure (prevents lungs being waterlogged).
- Sites at which rapid absorption occurs: in the small intestine and kidney, absorption into capillaries is the major function.

THE LYMPHATICS

The lymphatic system functions to carry filtered interstitial fluid back into the blood stream. It can be divided into two parts:

Initial lymphatics
These are made up of:
- A single layer of thin endothelial cells.
- Open gaps of 2–5 µm between cells.
- No valves.

A discontinuous basement membrane is attached by anchoring filaments to parenchymal cells. They contain no muscle cells, but actin-like filaments may be present. There is no evidence, however, that they are intrinsically contractile.

Collecting lymphatics
These have a smooth muscle intima and are actively contractile with an adrenergic innervation. In appearance, they consist of a ladder of compartments (called *lymphangions*) separated by bileaflet valves. They exhibit peristaltic-like contractions with synchronised opening and closing of valves, producing a pump-like action moving fluid against gravity.

Lymph transport
In the initial lymphatics it is thought that the pressure pulsations (from nearby arteries and arterioles) and tissue stresses (muscle contraction, respiratory movements, etc.) may influence the filling of these vessels and the transport to collecting lymphatics. The latter transport the lymph received by a peristaltic pump-like action.

Composition of lymph
Lymph has a composition which is similar in composition to that of the ISF for water and salts. It contains:

- A lower protein concentration than that of plasma.
- A protein concentration that is greater than that of the ISF, as water is absorbed by the lymph glands.
- About half the plasma protein which escapes from the plasma each day and is returned to the circulation through the lymphatics.
- A composition that is different in different regions of the body, e.g. protein content is higher from the gut and liver than from skeletal muscle.

Estimated net movement of lymph in adults per day

According to the Starling hypothesis, fluid movement has been calculated to be:

ultrafiltration (20 L) = osmosis (16–18 L) + lymph flow (2–4 L)
(arterial end)   (venous end)

These figures now have to be revised because they were based on the amount of lymph reaching the thoracic duct. It is now known that the lymph glands absorb considerable amounts of water, so underestimating the amount of lymph reaching the thoracic duct.

Figure 1.11 illustrates the formation of lymph.

Exchange of water between the ECF and ICF

Water distribution is determined by osmotic forces (number of osmotically active particles per unit weight of fluid):

$$ECF_{osm} = ICF_{osm}$$

Water moves freely through membranes between compartments. In ECF, Na^+ is the major osmotically active cation. In the ICF, K^+ is the osmotically active cation. Cation distribution is determined by transport processes in membranes (see p. 350) and Gibbs/Donnan equilibrium.

Oedema

Oedema is the accumulation of excessive amounts of fluid in the extracellular space. There are two types of oedema: general oedema and local oedema.

General oedema

When the intake of fluid becomes chronically greater than output, oedema will occur. Its distribution is determined by gravity.

The cardinal sign of oedema is pitting, i.e. firm and sustained pressure with the thumb over the ankle (or sacrum if in bed) causes a pit which is slow to refill. Pitting can only be demonstrated when the body weight increases by 10–15% due to fluid retention.

In the pulmonary circulation oedema can be detected by hearing moist crepitations (crackles) over the base of the lung with the stethoscope.

The primary causes of general oedema are:

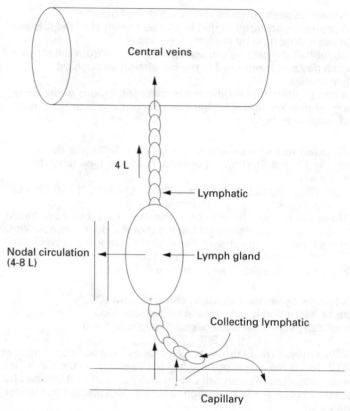

Fig. 1.11
The formation of lymph. Tissue fluid circulation (L/day).

1. *Cardiac failure*
   Increased central venous pressure increasing the mean P_c and formation of ICF
   Impaired renal blood flow→reduction in GFR→excessive re-absorption of filtrate.
2. *Hypoproteinaemia*
   Inadequate protein intake (famine; fad foods)
   Failure to digest protein (exocrine pancreatic failure)
   Failure to absorb protein
   Reduced synthesis of albumin (liver disease)
   Excessive protein loss (e.g. through the kidney – proteinuria).
3. *Renal causes*
   Renal disease
      protein loss
      reduced GFR.

Local oedema
Local oedema occurs in discrete regions or tissues. There are several causes of local oedema:
1. *Venous causes*
   Venous obstruction:
      pressure on veins
      venous thrombosis (clotting in a vessel)
2. *Lymphatic causes*
   Obstruction to lymphatic drainage:
      mechanical
      filarial worms
      tumour infiltration
3. *Allergic and inflammatory causes*
   Increase in capillary permeability→escape of proteins and fluids into the ISF.

REGULATION OF ARTERIAL BLOOD PRESSURE

THE 'NORMAL' BLOOD PRESSURE

The blood pressure is not constant but varies depending upon the conditions under which it is taken. It varies with:

Age
The mean blood pressure increases progressively with age, from about 100/65 mmHg in children, 130/85 mmHg at about 30 years to 180/90 mmHg at 70 years. Note that the pulse pressure gradually increases and this is related to a fall in vessel compliance (hardening of the arteries).

Exercise
With exercise the cardiac output rises but at the same time there is vasodilatation in the muscles and so the peripheral resistance falls. The blood pressure therefore depends upon the relative change in these parameters. The blood pressure can rise to quite high levels during heavy static exercise (weight lifting). After 10 min or so brisk walking the blood pressure can fall below the level it was at before exercise; this is the basis for recommending exercise in patients who are moderately hypertensive.

Sleep
During sleep blood pressure can fall to very low levels.

Emotion and stress
Anger, fear, excitement, etc. are stimuli which will raise the blood pressure. Apprehension on a visit to the doctor will elevate the blood pressure.

Position of the site of measurement
Blood pressure in the foot will be increased to the extent of the
height of the column of blood above the heart. Above the heart it
will be reduced.

Respiration
Smaller variations occur in phase with respiration.

SHORT-TERM REGULATION OF THE BLOOD PRESSURE

As the mean blood pressure is a function of the cardiac output and
the peripheral resistance, regulation is brought about by altering
one or both of these parameters. In the steady state the blood
pressure is held constant at a level which depends upon
information coming from a number of receptor inputs which act to
adjust these variables.

The short-term regulation of blood pressure is illustrated above.

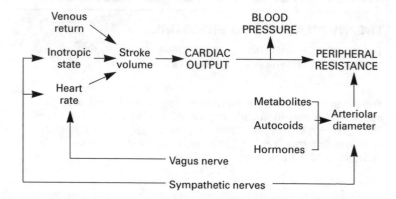

The principal mechanisms involved in reflex modulation involve:
- Sympathetic nerves which influence arteriolar diameter and the
  rate and force of contraction of the heart.
- Parasympathetic (vagal) nerves which influence the heart rate.
  The receptor inputs are:
- *Arterial baroreceptors* situated in the adventitia of the aortic arch
  and the carotid sinus.
- *Atrial stretch receptors* situated at the vena cava–right atrial and
  pulmonary venous–left arterial junction which monitors the
  heart's end diastolic volume.

Arterial baroreceptor reflexes
The components of these reflexes include:

Nerve endings   Free and encapsulated in the adventitia of the aortic arch and carotid sinus.

Nerve pathway   Afferent neurones travel from the aortic arch to the cardiovascular centres in the vagus nerves and from the carotid sinus in the glossopharyngeal nerve.

Stimuli   A rise in arterial pressure→increase in transmural pressure→stimulation of baroreceptor endings.

Efferent pathways

An increase in parasympathetic discharge→bradycardia (slow heart rate).

A decrease in sympathetic discharge:

   to the sino-aortic node→bradycardia

   to ventricular muscle→decrease in contractility→decrease in stroke volume

   to arterioles→decreased total peripheral resistance (greatest in splanchnic area)

   to the veins→increased venous compliance and capacity→reduced diastolic volume→fall in stroke volume.

These effects are in the opposite direction to the initial rise, thus restoring the blood pressure back towards normal. If the blood pressure falls then the converse effects occur, which is usually more frequent in medical practice.

Acute hypotension

This is corrected by mechanisms which:

- Stimulate cardiac output.
- Increase total peripheral resistance.
- Cause fluid retention to enhance the plasma volume.
  These mechanisms are:

1. *Tachycardia* – increase in heart rate.
2. *Venoconstriction* – particularly the splanchnics→transfers blood to central veins→enhances stroke volume.
3. *Vasoconstriction* to splanchics, renal circulation and skeletal muscle increases total peripheral resistance which leads to indirect effects on extracellular fluid volume as a consequence of vasoconstriction.
4. *Capillary pressure lowered* by arteriolar constriction→absorption of interstitial fluid→expansion of plasma volume.
5. *Increased renal nerve activity*→secretion of renin→secretion of angiotensin: which gives rise to a generalised contraction of vascular smooth muscle reducing the capacity of the vascular system and to the release of aldosterone from the adrenals→retention of Na$^+$ and water by the kidney.
6. *Release of vasopressin* from the posterior pituitary by a reduction of baroreceptor activity, leads to an anti-diuresis (reduction in the output of urine).

Cardiac and pulmonary receptors

Afferent fibres from these areas have, overall, a tonic inhibitory effect on heart rate and peripheral vascular tone when activated. There are three main functional classes of receptors:

Mechanoreceptors
Situated around the veno-atrial junctions→discharge into myelinated vagal fibres.
Scattered throughout atria, ventricles and pulmonary artery→discharge into non-myelinated fibres of vagus and cardiac sympathetics.

Chemoreceptors
These discharge into both vagus and cardiac sympathetics.

Atrial stretch reflexes

Atrial reflexes monitor acute changes in atrial volume and pressure and long-term changes in blood volume.

Receptors and afferent pathways
Discrete endings located in the endocardium at the junction of the great veins of both atria and attached to myelinated fibres which run in the vagus nerves to cardiovascular centres.

Efferent pathway
Cardiovascular centres→down spinal cord→cardiac sympathetics: cardiac sympathetics→sino-atrial node (not to cardiac muscle)
    activity in lumbar and splenic nerves unchanged
    activity in efferent renal nerves decreased.
    Thus atrial receptors are stimulated by increases in atrial volume and pressure. The reflex responses are:
- An increase in heart rate.
- A decrease in renal vascular resistance.
- An increase in urine flow, and in free water clearance and decreases in plasma vasopressin, cortisol and renin.

CARDIOVASCULAR CONTROL CENTRES

The neurone systems which control the heart and blood vessels can be found in at least five different sites in the central nervous system:
- Spinal cord.
- Medulla oblongata.
- Cerebellum.
- Hypothalamus.
- Cerebral cortex.

SPINAL CORD

Section through the spinal cord at a high level results in a steep fall in blood pressure, demonstrating that the normal tonic effect on the arterioles relies on impulses coming from higher centres. Over a period of days to weeks, neurones of the intermediary lateral columns regain some activity from impulses arriving at the cord from the periphery, allowing some improvement of arteriolar tone and blood pressure.

MEDULLA OBLONGATA

Much of the integration occurs in the medulla which receives information from the cardiovascular receptors (baroreceptors, cardiopulmonary receptors, etc.) and from which are two outputs, one through the vagus nerve, the other through the sympathetics via the spinal cord.

Vagal outflow
Afferent fibres from the IXth and Xth cranial nerves enter the medulla to synapse in the *nucleus of the tractus solitarius*. Neurones then pass from here to three areas:
* *The nucleus ambiguus* from which preganglionic vagal neurones arise.
* *The dorsal motor nucleus of the vagus* from which a smaller number of preganglionic vagal fibres arise.
* *The hypothalamus* (depressor area) from which fibres pass back to the dorsal vagal motor nucleus.

Sympathetic outflow
Preganglionic sympathetic fibres which supply arterioles, veins, myocardial muscle and the SA node, arise from the intermediolateral cell column of the spinal cord. These neurones are controlled by three groups of cells in the medulla which are:
* Raphe nucleus.
* Cells in the caudal ventrolateral group (caudal vasodepressor area).
* Cells in the rostral ventrolateral group (rostral vasopressor area).

The afferent neurones synapse in the nucleus of the tractus solitarius which sends fibres through polysynaptic pathways to the raphe nucleus, the caudal ventrolateral group of cells and the rostral ventrolateral group of cells. The rostral vasopressor cells are excitatory and are tonically active, but are modulated by γ-aminobutyric (GABA) inhibitory fibres from the caudal vasodepressor neurones. The raphe nucleus projects to the intermediolateral column in the spinal cord and exhibits an inhibitory influence there.

Afferent fibres from work receptors in muscle terminate in the lateral reticular nucleus and the nucleus of the tractus solitarius.

The principal neurone groups in the medulla controlling the heart and blood vessels are shown below.

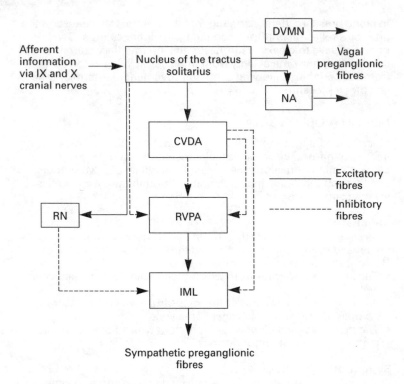

CVDA = caudal vasodepressor area
DVMN = dorsal vagal motor nucleus
IML   = intermediolateral cell column
NA   = nucleus ambiguus
RN   = raphe nucleus
RVPA = rostral vasopressor area

HYPOTHALAMUS

Regions involved in cardiovascular integration are the: depressor area; defence area; temperature-regulating area; the area controlling secretion of vasopressin (ADH).

Depressor area
Input from nucleus of the tractus solitarius (baroreceptor reflex) output activates the dorsal motor nucleus and the cardiovasodepressor area in the medulla.

Defence area (alerting)
This area is activated by the limbic system which results in:

- An increase in heart rate, cardiac output and blood pressure.
- Vasoconstriction, except in the heart, brain and skeletal musculature.
- In animals manifestations of fear and rage.
- An inhibitory effect on the nucleus ambiguus, caudal vasodepressor area and baroreceptor input to the nucleus of the tractus solitarius.

Temperature-regulating area
See page 333.

Area controlling the secretion of vasopressin
The supraoptic and paraventricular nuclei contain cells which produce vasopressin (ADH). The axons of these neurones transport ADH into the posterior pituitary. Activity of anterior pituitary nuclei comes from local osmoreceptors and nucleus of the tractus solitarius.

CEREBELLUM

The fastigial nucleus and vermal cortex receive projections from the medulla which not only take part in the coordination of muscular movement in exercise but also the accompanying cardiovascular response.

CEREBRAL CORTEX

The cerebral cortex can initiate many cardiovascular responses (just as it can muscular activity) by acting through the brain stem. The effects of excitement and anxiety, which have effects on heart rate and blood pressure, may emanate from the cerebral cortex, through the limbic system.

VASOMOTOR REFLEXES

These reflexes are activated as a result of shifts in blood volume brought about by postural change or othostasis.

Shifting blood away from the chest
Stimulus:
- Tilt into a foot-down position.
- Negative pressure applied to the lower part of the body.
- Positive pressure breathing.
- Valsalva manoeuvre.

Response to a passive tilt
During quiet standing for 15 min there is a fall in plasma volume

of some 18% ($\cong$ 700 ml). This results in a fall in stroke volume and cardiac output, with a decrease in pulse pressure. The mean blood pressure may fall but this is only transitory as reflex compensatory mechanisms come into play. These mechanisms are due to reflex effects from carotid and aortic baroreceptors, which are:

- Tachycardia and an inotropic effect (rise in myocardial contractility).
- Increased peripheral resistance due to reflex vasoconstriction in muscle, splanchnic and renal areas.
- Reflex venoconstriction (not sustained in muscle veins).

Changes in fluid-regulating hormones together with renal vasoconstriction lead to a reduction in electrolyte and water excretion by the kidneys. This involves reflex increases in vasopressin secretion and renin–angiotensin–aldosterone secretion.

Shift of blood towards the chest

Stimulus
Manoeuvres which increase intrathoracic fluid volumes: tilt to foot-up (head-down) position.
Application of positive pressure to the lower limbs.
The experience of astronauts as they enter weightlessness.
Immersion in water.

Response
In general a reversal of the effects of foot-down tilt.
Increase in thoracic volume.
Reflex vasodilatation.
Inhibition of vasopressin release (depends upon the state of hydration).
Increase in atrionaturetic peptide.
Suppression of plasma renin and aldosterone secretion.

THERMOREGULATORY REFLEXES

Afferent pathway
Temperature receptors in the skin.
Direct action of warm blood on the hypothalamic centres.

Centres
Hypothalamus.

Efferent pathways
Hypothalamus→medullary sympathetic centres→release of vasoconstrictor tone.
Sympathetic cholinergic action on sweat glands→liberation of bradykinin-forming enzyme.
Exposure to cold can cause a rise in blood pressure.

CHEMORECEPTOR REFLEXES

Arterial chemoreceptors

Receptors
Carotid and aortic bodies: highly vascular nodules near to the carotid sinus and aortic arch.

Afferent pathway
Travel with baroreceptor fibres in the IXth and Xth cranial nerves→medulla
They respond to:
* Hypoxia.
* Hypercapnia.
* Acidaemia.

Function
Mainly in respiratory control at normal blood gas concentrations, but during more severe hypoxia or hypercapnia→increased sympathetic vasoconstrictor drive (except skin). This reflex is important in:
* Asphyxia.
* Haemorrhage (severe).
* Hypotension→stagnant hypoxia→metabolic acidosis→sympathetic drive→increases blood pressure→counteracts hypotension and causes tachypnoea characteristic of a fall in blood pressure seen in severe haemorrhage.
  In hypercapnia the CO_2 excess reflexly excites vasoconstrictor fibres but locally causes vasodilatation.

REFLEXES FROM THE LUNG

Mechanoreceptors in the lung are stimulated on inspiration. This results in activation of afferent fibres in vagus which are directed towards the medulla. Activation of the reflexes causes sinus arrhythmia, which are most prominent in the young.
    The efferent pathways involved are:
Vagus→reduced vagal tone→tachycardia.
Sympathetic→reduced activity→vasodilatation.

EMOTIONAL STRESS REFLEXES

Emotional stress gives rise to intense vasodilatation in muscle and to vagal stimulation of the heart with a bradycardia. The causes include sympathetic cholinergic vasodilatation, probably mediated by the anterior hypothalamus and cerebral cortex. Adrenaline from the adrenal medulla also probably contributes. Then follows:
* A marked fall in arterial blood pressure.

- A reduced blood flow to the brain.
- Loss of consciousness, i.e. a faint.
  The condition is known as *vago-vagal syncope*.

OTHER EFFECTS ON THE CARDIOVASCULAR SYSTEM

Pain

Somatic pain→tachycardia and an increase in blood pressure (e.g. pinch of the skin).
Visceral pain→bradycardia and hypotension.
Bladder distension→tachycardia, vasoconstriction in the limbs and hypertension.

Humoral control

Action of injected adrenaline
Immediate and transient fall in arterial blood pressure, followed by a sudden rise which lasts but a few minutes.
Constriction of cutaneous vessels.
Dilatation of vessels in skeletal muscle.
Total peripheral resistance lowered.

Action of injected noradrenaline
Rise in arterial blood pressure not preceded by a fall.
Reduction in cardiac output through aortic and carotid sinus reflex.
Noradrenaline is a general vasoconstrictor.
Contained in nerve endings in walls or arteries, veins and heart.

Angiotensin

Octapeptide – angiotensin II.
Generalised vasoconstriction.
Stimulates the production of aldosterone.

Naturetic peptides

Composition 28 amino acids.
Stored in right atrium.
Infusion causes a fall in blood pressure.
Inhibits adenylate cyclase but activates guanylate cyclase.
Relaxes arterial walls.
Decreases the activity of the renin–angiotensin–aldosterone system.
Decreases angiotensin II production.
Direct inhibiting effect on the zona glomerulosa of the adrenal gland.

Kinins

Peptides with a powerful vasodilator action.
Bradykinin (mol. wt 1934).

Formed during active secretion of exocrine glands (viz. salivary glands).
Responsible for vasodilatation in many active tissues.

Histamine
Injected subcutaneously.
Vasodilatation of resistance vessels – causes headache.
Flushing of skin.
Fall in arterial pressure.
Increases permeability of capillaries.
Stimulates gastric acid secretion.

Antidromic impulses and vasodilatation
Stimulation of a posterior root (sensory) spinal nerve gives rise to vasodilation→axon reflex.

RESPONSE OF SKIN VESSELS TO MECHANICAL STIMULATION

Gentle stimulation with a blunt point leaves a white line (caused by capillary contraction). More vigorous stimulation activates the *triple response*. This involves:
1. Red line→capillary dilatation.
2. Flare→arteriolar dilatation.
3. Wheal→leak of plasma from capillary.

THE ELECTROCARDIOGRAM

The record of the electrical activity of the heart recorded from the surface of the body is known as the electrocardiogram. The instrument from which this recording is made is called the electrocardiograph. There are three sets of leads used to connect the body to the instrument:
1. The bipolar leads.
2. The unipolar leads.
3. The precordial or chest leads.
   In connecting the leads two important facts must be borne in mind:
1. The trunk behaves as a *volume conductor* and therefore the site at which the electrodes are attached affects the record obtained.
2. The limbs behave as *linear conductors*, and so it does not matter where on a limb (up to the shoulder or groin) the electrodes are attached. They behave as extensions to the wire connections.
   Einthoven made a number of simplifying assumptions, the most important being that the heart is at the centre of an equilateral triangle (base between the left and right shoulders and the apex in the left groin), and that it behaves as a dipole (Fig. 1.12).

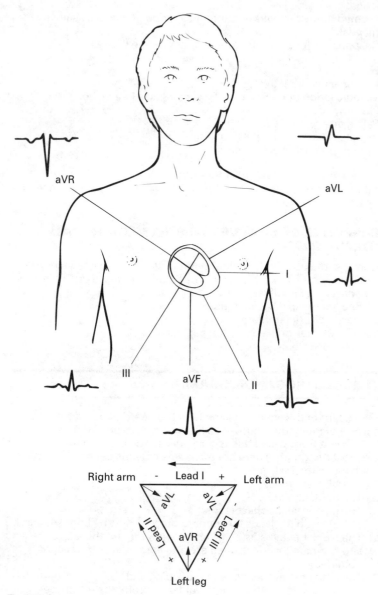

Fig. 1.12
Einthoven's triangle. For a definition of the recording lead connections, see the text.

Recording lead connections

Bipolar leads
I = Left arm – right arm.
II = Left leg – right arm.
III = Left leg – left arm.

Unipolar leads
aVL = Left arm – indifferent electrode (right arm + left leg).
aVR = Right arm – indifferent electrode (left arm + left leg).
aVL = Left leg – indifferent electrode (left arm + right arm).
 Figure 1.13 illustrates the position of the heart in the chest.

Chest leads (precordial)
There are six leads numbered V1 through to V6.
VI – right sternal margin, 4th space.
V2 – left sternal margin, 4th space.
V3 – midway between V2 and V4.

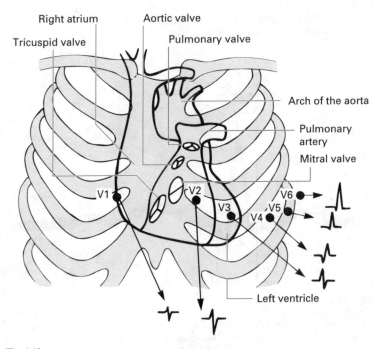

Fig. 1.13
The position of the heart in the chest and the site of the openings of the valves in relation to the surface of the chest. The position of the precordial ECG leads with typical records.

V4 – at midclavicular line at 5th space.
V5 – at left anterior axillary line and a line horizontally through V4.
V6 – at left midclavicular line and a line drawn horizontally through V4 and V5.

A record of an ECG

The components of the ECG (Fig. 1.14) are lettered for ease of description, P, Q, R, S and T waves:

- P wave is atrial activation – duration 60–110 ms.
- PR interval – 120–200 ms – measures the time taken for:
  (i) passage of atrial depolarisation wave to atrioventricular node (AV node);
  (ii) the delay imposed at the AV node by slower conducting tissue;
  (iii) the rapid spread through the bundle of His and branches to start ventricular depolarisation;
  most time is spent at stage ii.
- QRS complex is ventricular activation and lasts 60–110 ms.
- The QT interval is inversely related to the heart rate.
- ST interval – all parts of the ventricles depolarised.
- T wave is ventricular repolarisation.

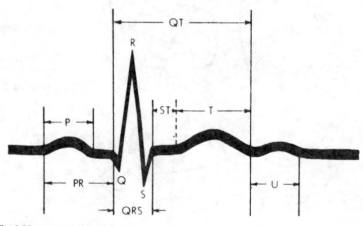

Fig. 1.14
The components of the electrocardiogram.
P wave – duration 60–110 ms.
PR interval – 120–200 ms, measures time taken for:
 (i) passage of atrial depolarisation wave to AV node;
(ii) delay imposed at AV node by slower conducting tissue;
(iii) rapid spread through AV node and bundle to start of ventricular depolarisation;
    most of the time is spent at stage (ii).

The view of the heart taken by the electrodes

For ease in interpreting ECG records it is best to consider the directions from which the various leads 'look' at the heart:

- Leads aVL and lead I 'look' at the heart from the anterolateral aspect.
- Leads II, aVF, and III 'look' at the heart from an inferolateral aspect.
- Lead aVR 'looks' into the cavity of the heart.

Shape of the QRS complex in the limb leads

Normal heart

When a wave of depolarisation spreads towards a lead, then the deflection of the ECG pointer is upwards (R>S). When the spread is away from the lead then the deflection is downwards (S>R). The wave of depolarisation spreads across the heart in many directions, but through the ventricles there is an average (or principal) direction. In the normal heart this is towards Lead II. So that in:

- Leads I, II, III and aVL the deflection will be upwards (R).
- Lead aVR the deflection will be downwards (S) – vector moving away.
- Lead aVL the deflection R and S will be approximately equal – at right angles.

Heart with hypertrophied right ventricle

This would result from chronic chest disease with an increased peripheral resistance in the pulmonary circulation. This is called right axis deviation, i.e. the electrical axis has swung away from Lead II (because of the greater mass of the right ventricle) towards Lead III. In this case Leads II, III and aVF would have a positive deflection (R), whereas Leads aVR, aVL and I would have negative deflections (C).

Shape of the QRS complex in the chest leads

With regard to the chest leads they 'look' at the heart in the horizontal plane:

- V1 and V2 'look' at events in the right ventricle.
- V3 and V4 'look' at events in the septum.
- V5 and V6 'look' at events in the anterior and lateral walls of the left ventricle.

The QRS complex shows a steady progression from V1, where it is predominately downwards (negative), to V6 where it is predominately upwards (positive). The transition point where R and S are equal indicates the position of the I-V septum.

Information obtained from an ECG recording

The ECG can give information on:

- Heart rate.
- Rhythm (regular or irregular).

- Relative increase in muscle mass.
- Damage to different parts of the heart.
- Disturbances of electrolyte concentrations in the blood, e.g. Ca^{2+} and K^+.
- Inflammation of the pericardium and the presence of fluid in the pericardium.

2. The autonomic nervous system

It is now recognised that the autonomic nervous system is composed of three divisions:
- The sympathetic nervous system.
- The parasympathetic nervous system.
- The enteric nervous system.

All the systems have both motor and sensory nerves mixed together.

THE MOTOR OUTFLOW OF THE AUTONOMIC NERVOUS SYSTEM

Motor outflow arises from the intermediary lateral cell column of the thoracic and parts of the lumbar spinal cord (T1–T12 and L1–L2). The fibres leave the spinal cord at the same segment at which the cell body lies (the preganglionic fibres), into the ventral root. From there they pass out into white rami communicans (myelinated fibres) and then travel to:
- Sympathetic paravertebral ganglia including the stellate ganglion, middle and superior cervical ganglia.
- Prevertebral (preaortic) ganglia – coeliac, superior and inferior mesenteric ganglia.

The relay of information occurs in these ganglia and fibres, the postganglionic neurones, leave to innervate organs. These are non-myelinated and appear grey. They reach their target organs through two routes:
1. Along spinal nerves which they join as grey rami communicans, e.g. to skin vessels.
2. Along the blood vessels which supply the target organ, e.g. coelic axis and branches to the stomach and small intestine.

Note that the adrenal medulla is supplied only by preganglionic fibres, as the medulla itself is derived from the same cells as the postganglionic fibres.

NEUROTRANSMITTERS IN THE SYMPATHETIC NERVOUS SYSTEM

The classical description of chemical transmitters in the autonomic nervous system is that the preganglionic neurone is cholinergic and the postganglionic fibre is adrenergic (noradrenaline). The exceptions are those to the sweat glands and skeletal muscle blood vessels, which are cholinergic. Recent work shows that there is more than one postganglionic transmitter and that there may be co-localisation of transmitters.

Adrenal medulla

The adrenal medulla releases adrenaline (80%) and noradrenaline (20%). Noradrenaline is converted to adrenaline by the enzyme phenylethanolamine N-methyltransferase (PNMT).

Noradrenaline ───────→ adrenaline
                (PNMT)

PNMT is not present in nerves and therefore there is no adrenaline released from nerve terminals.

The distribution of the sympathetic nerve fibres is shown in Figure 2.1.

Comparison of effects of adrenaline and noradrenaline in man

	L-adrenaline	L-noradrenaline
Heart rate	Increased	Decreased
Cardiac output	Increased	Variable
Total peripheral resistance	Decreased	Increased
Blood pressure	Rise	Greater rise
Respiration	Stimulation	Stimulation
Skin vessels	Constriction	Constriction
Muscle vessels	Dilatation	Constriction
Bronchus	Dilatation	Less dilatation
Eosinophil count	Increased	No effect
Metabolism	Increased	Slight effect
O_2 consumption	Increased	No effect
Blood sugar	Increased	Slight increase
CNS	Anxiety	No effect
Uterus in late pregnancy	Inhibits	Stimulates
Kidney	Vasoconstrictor	Vasoconstrictor

(Reproduced with permission from: Bell GH, Emslie-Smith D, Paterson C 1980 Textbook of physiology, 10th edn. Edinburgh: Churchill Livingstone, p 401.)

Adrenergic receptors

Adrenaline and noradrenaline produce their effects by reacting with receptors which are of two general types:
- α-receptors.
- β-receptors.

These have been subdivided as follows:

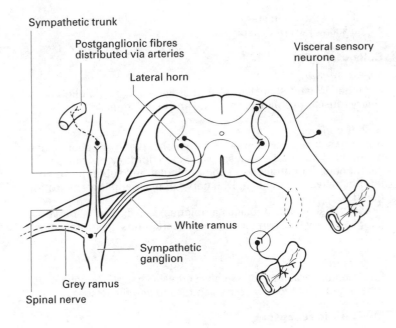

Fig. 2.1
Origin and distribution of sympathetic neurones.

Adrenoreceptors by type and agonist potency

Type	α_{1a}	α_{1b}	α_{1c}	α_{2a}	α_{2b}
Potency	Nad > Ad	Nad = Ad	Nad = Ad	Ad > Nad	Ad > Nad
Type	β_1	β_2	β_3		
Potency	Nad > Ad	Ad > Nad	Nad > Ad		

Cellular action

Adrenaline and noradrenaline act on cells, after combining with their receptor, through second messengers:

- For α_{1a}, α_{1b} and α_{1c} the second messenger is inositol triphosphate/diacylglycerol.
- For α_{2a} and α_{2b} the second messenger is a fall in cAMP.
- For β_1, β_2 and β_3 the second messenger is a rise in cAMP.

THE PARASYMPATHETIC NERVOUS SYSTEM

The parasympathetic outflow is from:

- Nuclei of the brain stem.
- The sacral part of the spinal cord.

Cells of origin

Tectal outflow
Endinger–Westphal nucleus → IIIrd nerve → ciliary ganglion → ciliary muscle → sphincter pupillae and ciliary muscle.

Bulbar outflow
Superior salivary nucleus → VIIth cranial nerve → sphenopalatine ganglion → lacrimal and nasal glands → chorda tympani → submandibular ganglion → submandibular and sublingual glands.
Inferior salivary nucleus → IXth nerve → otic ganglion → parotid gland.
Dorsal vagal nucleus (nucleus ambiguus) → Xth (vagus) nerve → ganglia innervated lying in organ → heart, lungs, gastrointestinal tract.

Sacral outflow
Intermediolateral nucleus → pelvic nerve → ganglia lying in organ innervated → bladder, rectum, genital and erectile tissues.

Muscarinic receptors

Type	M_1	M_2	M_3
Tissue	neural	cardiac	smooth muscle glands

Cellular action
After acetylcholine has combined with its receptor, its action is exerted through second messengers:
- For M_1 and M_3 the second messenger is inositol triphosphate/diacylglycerol.
- For M_2 the second messenger is a reduction in concentration of cAMP.

ENTERIC NERVOUS SYSTEM (ENS)

The enteric nervous system is made up of two sheets of neurones:
- The *myenteric plexus* (Auerbach's) lying between the inner circular layer and outer longitudinal smooth muscle layers.
- The *submucosal plexus* (Meissner's) lying within the submucosa.
  The ENS is a complex network of neurones, with an input from the parasympathetic and sympathetic divisions of the ANS, and from various afferent fibres within the gut wall. It has an output to the smooth muscles and glands of the gastrointestinal tract. The major functions of the ENS are to regulate motility, ion and water transport and it has a role in the secretion of enzymes.
  The ENS can regulate all the basic functions of the gastrointestinal tract in the absence of any external innervation.

This, however, does not negate the role of the external nerve supply. Within the ENS are many diverse functional types of neurone and many transmitters have been described, some proven, others putative. The following (in addition to noradrenaline and acetylcholine) have been implicated, either as transmitters or modulators:

Adenosine and purine nucleotides
γ-Aminobutyric acid (GABA)
Vasoactive intestinal polypeptide
Somatostatin
Dynorphins and encephalins
Gastrin-releasing peptide
Galanin
Inter alia

Serotonin (5HT)
Bradykinin
Substance P
Cholecystokinin (CCH-8)
Bombesin
Neuropeptide Y
Calcitonin gene-related peptide

Most of these have been identified in nerve or nerve endings by immunoreactive methods and some may be co-localised in the same cells as noradrenaline, but most have not met all the criteria listed below:

The criteria from which chemical substances are established as neurotransmitters
1. There must be morphological evidence for its localisation in the presynaptic neurone.
2. Processes for its biochemical synthesis, release and degradation are present.
3. Experimental application of the chemical fully mimics its synaptic action.

THE SENSORY COMPONENT OF THE AUTONOMIC NERVOUS SYSTEM

Nerves of both the sympathetic and parasympathetic systems carry afferent neurones and send back information to higher centres. The receptors do not show morphological specialisation and the information signalled depends upon the site of the receptor:
- Physical and chemical nature of luminal contents – receptors lying immediately below the mucosal epithelium.
- Changes in muscle tension – receptors lying in the muscle layers.
- Movement and distortion of the viscera – receptors in the mesenteric attachments.
- Composition of digestive products – receptors in the liver (osmolality, ionic concentration, glucose, temperature, pressure).

Central connections of the Xth and IXth nerves

The fibres follow the same route as the efferent neurones. Their cell bodies lie in the superior and inferior vagal ganglia (jugular and nodose) and the central connections end in the nucleus of the tractus solitarius (NTS). The projections from the NTS involve:

- *Reflex connections* with other brain stem nuclei giving rise to viscero-visceral reflexes through the dorsal vagal nucleus and nucleus ambiguus (see vago-vagal reflexes in the control of motility and secretion by stomach, Ch. 4). In the case of the glossopharyngeal nerve the cell bodies lie in the IXth ganglion.
- *Reflex taste responses* → lingual nerve → salivary nuclei → chorda tympani → submandibular ganglion → lingual and submandibular glands (see Ch. 4).
- *Ascending fibres* → midbrain → reticular nuclei → hypothalamus and cerebral cortex, e.g. in the case of taste fibres these pass via the IXth nerve → rostral part of NTS → pontine gustatory area → medial part of the ventral posterolateral area of thalamus → *insular* cerebral cortex.

Central connections of afferent splanchnic and pelvic nerves

The cell bodies in dorsal root ganglia. The central processes project to laminae I and V of the dorsal horn (viscerosomatic convergence occurs in dorsal horn). This is the basis of the theory of referred pain. The fibres then ascend within the spinal cord to the medulla in the vagal nucleus (splancho-vagal inhibition of stomach motility).

Fibres from hind gut and bladder

Second-order neurones give rise to projections extending to various segments. There are long ascending projections to higher centres. These fibres signal sensations relating to the bowel and bladder.

There are also somatic fields, i.e. perineal from pelvic nerve. They are associated with short ascending fibres involved in thoracolumbar intestino-intestinal inhibitory reflexes. These are local to the same or adjacent segments or descend to lower levels.

Cortical representation of autonomic function

At least three cortical areas play a role in autonomic function:

Insular area   This area contains an organotopic visceral sensory map (the best established).

Infralimbic cortex → descending projections to central autonomic control system. This area can be regarded as an autonomic motor area. The prelimbic area can be regarded as an autonomic premotor area.

The primary sensory and motor areas of the cortex   receive visceral afferents and activate autonomic responses. These are coordinated with ongoing somatic sensory and motor processes.

3. The respiratory system

Examples of symbols used in respiratory physiology

Symbols	Examples
V = gas volume	V_A = Volume of alveolar gas
$\dot{V}$ = gas volume/unit time	$\dot{V}O_2$ = O_2 consumption/minute
P = gas pressure	PA_{O_2} = alveolar O_2 pressure
F = fractional concentration in dry gas phase	$F_{I_{O_2}}$ = fractional concentration of O_2 in inspired gas phase
I = inspired	$F_{I_{CO_2}}$ = fractional concentration of CO_2
E = expired	V_E = volume of expired gas
A = alveolar	V_A = alveolar ventilation
T = tidal	V_I = tidal volume
D = dead space	V_D = volume of gas in dead space
B = barometric	P_P = barometric pressure
Q = volume of blood	
$\dot{Q}$ = volume of blood per unit time	
a = arterial	
v = venous	
c = capillary	

THE NOSE AND NASAL PASSAGES

A vascular mucous membrane, covers the nasal septum and the conchas. It consists of an epithelial layer of ciliated columnar cells intermixed with goblet cells. In the lamina propria there are mucous and serous glands.

The epithelium functions are protective:

- Warm and moisten inhaled air.
- Remove dust particles.
- Cilia move mucus towards the nasopharynx.
  Irritation of the mucosa provokes:
- Apnoea.
- Closure of the larynx.
- Constriction of the bronchi.
- Bradycardia.
- Variable effects on the blood pressure.

Hyperaemia of the mucosa leads to congestion which can increase the resistance to air flow. The causes of hyperaemia are:
- Chemical.
- Mechanical.
- Infrared rays on the face.
- Infections.

THE TRACHEA AND BRONCHIOLES

The major air passages start at the trachea, which is a tube about 10 cm long and 2 cm in diameter. It is supported by U-shaped rings of cartilage connected together posteriorly by smooth muscle. Descending into the lung the tube bifurcates into right and left bronchi with further branching into small bronchi (11 branchings). The branches, from 12 to 16, are called bronchioles (they lack cartilage). Branches 16–18 or 19 are the respiratory bronchioles. These lead into the alveolar ducts, then the alveolar sacs and from there into the alveoli.

BRONCHIAL MUSCLE

Bronchial smooth muscle is capable of causing contraction or relaxation.

Bronchoconstriction
Vagal stimulation
Cholinergic drugs
PGF_2 – ? through cholinergic receptors
Histamine (in asthmatics)
During expiration
Asphyxia.

Bronchodilatation
Sympathetic stimulation
Adrenaline – activates β_2-receptors
Vagotomy
Atropine
Inspiration
Inhaling $PGE_2 \rightarrow$ activates β-receptors.

BRONCHOSECRETION

Airway mucus is derived from several cell types. In the submucosal glands there are two types of cells in separate acini:
- Serous.
- Mucous.
  The acini are innervated by both cholinergic and adrenergic

neurones. The glands discharge into ducts that open on the epithelial surface, there the secretions partially mix with those from the epithelial cells forming 'airway surface liquid'.

Stimulation of airway gland secretion occurs by:

- Local action of mediators.
- Axon reflexes.

Motor pathways are activated reflexly from the central nervous system. Irritation in many sites of the respiratory tract (nose to trachea and bronchi) reflexly cause mucous secretion via receptors that are also responsible for sneeze and cough. Stimulation of peripheral chemoreceptors by hypoxia causes reflex secretion. As reflex secretion is largely inhibited by atropine (in humans), reflex mucous secretion appears to be largely cholinergic.

Adrenergic and non-adrenergic non-cholinergic responses also occur, but the action of modulators (neuropeptides, catecholamines, prostaglandins, etc.) varies from species to species.

FUNCTION OF AIRWAY MUCUS

The airway mucus functions to traps bacteria and collect cell debris. These are then removed by ciliary action towards the pharynx. Overproduction of airway mucus and its expectoration is one of the commonest symptoms of pulmonary disease.

ALVEOLAR–CAPILLARY MEMBRANE (BLOOD–GAS BARRIER)

Gas exchanges occur across this membrane which has a total surface area of about 70–90 m^2 (containing about 300 million alveoli) with a thickness of about 1 μm. To reach a red cell the oxygen molecule must pass through the following layers:

A thin layer of fluid lining the alveolus containing surfactant

Alveolar epithelium
Alveolar basement membrane
Interstitial space
} these three structures appear fused to form endothelial basement membrane, which is a single layer

Capillary endothelium

ALVEOLI

The lung can be regarded as being made up of 300 million bubbles (the alveoli). These vary in size, and from Laplace's law the pressure in an alveolus would be:

$$\text{Alveolar pressure} = \frac{2 \times \text{Surface tension}}{\text{Radius}}$$

If the surface tension was equal in all alveoli those with the smallest radius, having the greatest pressure, would empty into the large alveoli and therefore collapse. This does not happen because the surface tension is reduced in proportion to the radius. The alveoli are lined with a surfactant which reduces the surface tension and whose molecules pack more densely in the smaller alveoli.

Surfactant

Surfactant is a lipoprotein. It contains the phospholipid dipalmitoyl lecithin. It is synthesised by alveolar type II cells, from fatty acids, and extruded on to the alveoli. Secretion is not established until 28–30 weeks in utero and is stimulated by alveolar expansion and β-adrenergic mechanisms. The turnover of surfactant is very rapid (half-life about 2 h). Its main functions are to:

- Increase lung compliance.
- Stabilise the alveoli.
- Help to keep the postnatal lung dry.

Respiratory distress syndrome in premature neonates is probably due to lack of surfactant.

VENTILATION OF THE LUNGS

Ventilation is the movement of air into and out of the lungs. As the thorax enlarges, the lungs expand. During normal quiet breathing inspiration is active. The respiratory muscles contract, moving the chest wall away from its equilibrium position. This is caused by the elastic properties of the lung and chest wall.

The muscles involved are:

- Diaphragm – descends with the dome becoming flattened – and accounts for three-quarters of the tidal volume.
- External intercostal muscles – elevate the rib margins upward and outwards (bucket and pump handle) to increase all diameters of the chest.

Expiration is passive, involving the chest wall and lungs recoiling to an equilibrium position, which is 50–60% of the fully expanded state.

DEEP BREATHING AND DYSPNOEA

Inspiration
Diaphragm
External intercostals
Accessory muscle, scaleni, sternomastoid and trapezius.

Expiration
Contraction of:
    abdominal muscles
    internal intercostals.

EXPANSION OF THE LUNGS

The lung is covered with a membrane, the *visceral pleura*, and the inner surface of the chest by the *parietal pleura*. The two membranes hold the lung and the chest wall together. Between the two is a thin film of liquid which provides a cohesive force which stops the two from separating. If air is introduced into the pleural space (*pneumothorax*), the lungs do not remain attached to the thoracic wall but collapse due to an elastic recoil.

When breathing in takes place the chest enlarges and the lungs attached to the chest wall by the visceral pleura also enlarge, creating a pressure difference between atmospheric pressure and the alveoli, so that air is drawn into the lungs. When the inspiratory muscles relax the elastic recoil of both lungs and chest wall drive the air out.

INTRAPLEURAL PRESSURE

This is measured by passing a needle into the intrapleural space, attached to a manometer. Alternatively, a tube (with a long narrow balloon attached) is passed down into the oesophagus (which transmits pressure from the intrapleural space) and attached to a manometer. The pressure recorded here approximates to the intrapleural pressure.

	Intrapleural pressure	
	Inspiration	**Expiration**
Quiet ventilation	0 to -4 cmH$_2$O	1 cmH$_2$O
Forced ventilation	-20 to -40 cmH$_2$O	$+30$ cmH$_2$O

EVENTS ASSOCIATED WITH INSPIRATION

During inspiration:
Vocal folds move apart
Bronchi dilate
Bronchial tree elongates
Alveoli become wider
Intrathoracic pressure falls
Venous return to the heart increases
Right ventricular output increases
Blood in the pulmonary circulation increases.

ELASTIC PROPERTIES OF THE LUNGS AND CHEST

The elastic recoil force:

- Lungs plus chest is given by the pressure at the mouth (airways not opened to the atmosphere).
- Chest alone by the pleural pressure.
- Lungs alone by the transpulmonary pressure (mouth pressure minus pleural pressure).

- The compliance of each structure (ease with which it stretches) is measured by the change in volume divided by the change in pressure.

Compliance

Lung compliance (C_L) is defined as the change in volume of the lung ($\triangle V$) produced per unit distending pressure ($\triangle P_{tp}$).

$$C = V/P$$

where the transpulmonary pressure (P_{tp}) = alveolar pressure (P_{alv}) minus the intrapleural pressure (P_{ip}), i.e. $P_{tp} = P_{alv} - P_{ip}$.

Measurements are made under static conditions, i.e. no air flow in or out of the lung. The alveolar pressure must therefore be the same as the atmospheric pressure (which by convention is considered to be zero).

Thus the static lung compliance is:

$$C_L = \triangle V/\triangle P_{ip}$$

Lung volume is measured by spirometer and intrapleural pressure from the oesophagus.

	Compliance (L/cmH$_2$O)
Adult male	0.09–0.26
Newborn child	0.005
10-year-old child	0.06

It is customary to normalise compliance to take into account lung size, i.e. the *specific compliance*.

$$\text{Specific compliance} = \frac{\text{Measured compliance}}{\text{Lung volume at which it is measured}}$$

It is usual to take the subject's FRC as the reference lung volume.

Causes of lung recoil

The lung recoils as a consequence of elastin and collagen fibres in the bronchi and alveolar walls. The surface tension at the interface between gas in the alveoli and the liquid surface of the alveolar wall also contributes.

Breath sounds

Vesicular
Bronchial.

LUNG VOLUMES (Fig. 3.1)

All the following volumes given in the examples are for a 70-kg man. For an individual the volumes depend upon age, gender, body mass and physical fitness.

Resting tidal volume (V_t)

The volume of air breathed in and out in a single breath during quiet breathing (approx. 500 ml).

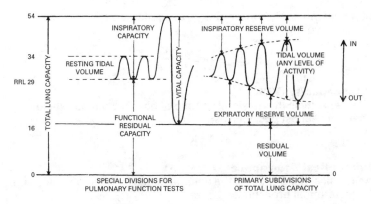

Fig. 3.1
Lung volumes. (Reproduced with permission from: Federation Proceedings 1950 9: 602.)

Tidal volume
The volume breathed in and out in a single breath at any level of activity.

Inspiratory capacity (IC)
The maximum volume of air that can be inhaled from the end of normal expiration.

Expiratory reserve volume (ERV)
The maximum volume of air that can be expired after a quiet expiration (approx. 1300 ml).

Residual volume (RV)
The volume of gas that remains in the lungs at the end of maximal expiration (approx. 1600 ml).

Functional residual capacity (FRC)
FRC cannot be measured by simple spirometry, because the lungs cannot be emptied completely by a maximal exhalation (approx. 29 000 ml).

$$FRC = ERV + RV$$

FRC can be measured using a helium dilution method:
A subject is connected to a spirometer containing a known quantity of helium.
After some breaths in and out the helium in the spirometer and the lungs are at equilibrium, so that:

The amount of He before equilibrium = $C_1 \times V_1$
The amount after equilibrium = $C_2 \times (V_1 + V_2)$

where:
C_1 = initial spirometer He concentration
V_1 = the initial spirometer volume

C_2 = final mixed He concentration (at equilibrium)
V_2 = lung volume

$$C_1 \times V_1 = C_2 \times (V_1 + V_2)$$
$$V_2 = V_1 (C_1 - C_2)/C_2.$$

Vital capacity (VC)
Vital capacity is the maximal volume of gas that can be exhaled after the deepest possible inspiration.

$$VC = IC + ERV$$

Anatomical dead space
Anatomical dead space is the volume of gas in the air passages up to the terminal bronchioles.

$$\text{Anatomical dead space} = \text{tidal volume} \times \frac{(F_{A_{CO_2}} - F_{E_{CO_2}})}{(F_{A_{CO_2}})}$$

where:
$F_{A_{CO_2}}$ = fractional composition of end expired CO_2
$F_{A_{CO_2}}$ = fractional composition of mixed expired CO_2

Physiological dead space
Physiological dead space is that part of the tidal volume which does not participate in gaseous exchange, as the P_{CO_2} of perfused alveoli is in equilibrium with the P_{CO_2} of arterial blood. Then:

$$\text{Physiological dead space} = \text{tidal volume} \times \frac{Pa_{CO_2} - PE_{CO_2}}{PA_{CO_2}}$$

where:
Pa_{O_2} = partial pressure of arterial CO_2
$Pa_{E_{CO_2}}$ = partial pressure of mixed expired CO_2
  Of the tidal volume breathed at rest, about two-thirds reaches the alveoli, this is the *alveolar volume* and is available for gas exchange. The remaining third, the anatomical dead space, fills the airways and part of the alveoli, the *alveolar dead space*, where the gases do not reach the alveolar capillaries.

$$\text{Physiological dead space} = \text{anatomical dead space plus the alveolar dead space}$$

Minute ventilation
Minute ventilation is the product of the tidal volume and the number of breaths per minute. About 6 L/min at rest and about 100 L/min during exercise.

Alveolar ventilation
Alveolar ventilation can be obtained by dividing the CO_2 output by the alveolar concentration of the gas.

$$V_A = V_{CO_2}/F_{A_{CO_2}}$$

where:
V_{CO_2} = the volume of CO_2 exhaled in unit time
$F_{A_{CO_2}}$ = the fractional concentration of CO_2 in alveolar gas.

DYNAMIC TESTS OF VENTILATION

Maximal breathing capacity (MBC)

Maximal breathing capacity (MBC) is the greatest volume of air which can be breathed by a subject per minute (measured over a 10 s period), approx. 150 L/min.

Airway resistance

This is measured by the ratio of the forced expiratory volume in 1 s and the forced vital capacity.

Forced expiratory volume in 1 second (FEV_1)

Forced expiratory volume in one second (FEV_1) is the volume of gas expired during the first second of forced vital capacity (FVC). The ratio FEV_1/FVC = 75% or more for the normal lung.

THE GAS LAWS

CHARLES' LAW

At constant pressure the volume of a gas is directly proportional to its absolute temperature.

BOYLE'S LAW

At constant temperature the volume of a gas is inversely related to its pressure.

$$P_1V_1 = P_2V_2$$

AVAGADRO'S LAW

Equal volumes of gas at the same temperature and pressure contain the same number of molecules.

GENERAL GAS LAW

$$PV = nRT$$

where:
P = pressure
V = volume
n = number of moles of gas
T = temperature in degrees Kelvin
R = the gas constant

where: P is in atmospheres, V is in litres and T is in °K then $R = 0.82$ L. atmosphere/M degree.

GRAHAM'S LAW OF DIFFUSION

The rate of diffusion of a gas is inversely proportional to the square root of its density.

DALTON'S LAW OF PARTIAL PRESSURES

In a mixture of gases, each gas exerts the pressure which it would exert if it occupied that volume alone, i.e. the total pressure (P_B) exerted by a gas mixture is the arithmetic sum of the partial pressures of the gases making up the mixture, e.g. for dry air:

$$P_B = P_{CO_2} + P_{O_2} + P_{N_2}$$

For dry atmospheric air:

$$P_{O_2} = 20.93 \times 760/100 = 159 \text{ mmHg}$$
$$P_{CO_2} = 0.04 \times 760/100 = 0.3 \text{ mmHg}$$
$$P_{N_2} = 79.03 \times 760/100 = 601 \text{ mmHg}.$$

HENRY'S LAW

The volume of a gas dissolved in a liquid is proportional to its partial pressure.

Gas dissolved = αP_{gas} (at constant temperature)
(α = solubility coefficient)

TENSION OR PARTIAL PRESSURE OF A GAS IN A LIQUID

At a given temperature, where no more gas dissolves in the liquid, saturation has occurred and an equilibrium exists between the gaseous and liquid phases, i.e. as many molecules of the gas enter the liquid as leave it. The gas in the solution then exerts the same tension as the partial pressure of the gas above the liquid.

Volumes are usually measured under ATPS conditions but should be stated in units of BTPS for ventilation and STPS for gas exchange.

ATPS = ambient temperature, pressure – saturated with water
STPD = standard temperature and pressure-dry (°C, 760 mmHg)
BTPS = body temperature and pressure – saturated with water.

ALVEOLAR GAS PRESSURES

Only that part of the minute ventilation that reaches the alveoli takes part in gas exchange.

O_2 is absorbed into the pulmonary capillary blood and CO_2 passes back into the alveolar gas.

% composition dry gas	Inspired air	Expired air	Alveolar air
O_2	20.93	16.89	14.50
N_2	79.04	79.61	79.95
CO_2	0.03	3.50	5.55

ALVEOLAR GAS PARTIAL PRESSURES (dry)

Oxygen

$$P_{A_{O_2}} = F_{A_{O_2}} \times (P_B - 47)$$
$$= 14.5 \times (760 - 47)/100$$
$$= 14.5 \times 713/100$$
$$= 105 \text{ mmHg}$$

Carbon dioxide

$$P_{A_{CO_2}} = F_{A_{CO_2}} \times (P_B - 47)$$
$$= 5.6 \times 713/100$$
$$= 40 \text{ mmHg}$$

The normal mean alveolar P_{O_2} and P_{CO_2} are determined by the rate of ventilation (flushing the alveoli with air) of the alveoli and the rate of metabolism (production of CO_2). As the CO_2 content of inspired air is virtually zero then:

Amount of CO_2 produced/min = Amount cleared from the alveoli by exhalation/min

$$\dot{V}_{CO_2} = V_A \times F_{A_{CO_2}}$$
$$F_{A_{CO_2}} = \dot{V}_{CO_2}/V_A \text{ or related to partial pressures}$$
$$P_{A_{CO_2}}/P_B - 47 = \dot{V}_{CO_2}/V_A$$

Thus for a given rate of CO_2 production the steady state partial pressure of alveolar carbon dioxide is inversely related to the rate of alveolar ventilation.

GAS EXCHANGE

In an adult:
- The resting oxygen consumption is about 250–300 ml/min.
- The resting carbon dioxide production is about 200–250 ml/min.

$$\text{Respiratory exchange ratio} = \frac{\text{Volume of carbon dioxide produced}}{\text{Volume of oxygen consumed}}$$

This ratio, also known as the *respiratory quotient*, depends upon the chemical nature of the food consumed (see p. 158).

Ventilatory requirement

The ratio of ventilation (ventilatory equivalent) to oxygen consumption is:

$$\dot{V}_E/\dot{V}_{O_2} \ (7.5/0.3 = 25 \text{ L/min})$$

Oxygen extraction coefficient

This is the amount of oxygen used divided by the amount inspired, expressed as a percentage $\dot{V}_{O_2}/(V_I \times F_{I_{O_2}}) \times 100$. At rest this is about 15–20%.

THE TRANSFER OF GASES AT THE ALVEOLAR–CAPILLARY MEMBRANES

By the time air has reached the alveolar ducts on inspiration, its velocity has been reduced to a very low value and it has been calculated that gas movement from the alveolar duct to the alveolar surface membrane is accomplished by diffusion. The rate of gas exchange depends upon:

- Partial pressure gradient of the gas.
- Solubility of the gas in cellular and liquid phases of the barrier.
- Area and thickness of the barrier (see p. 77).
- Diffusion constant (d) of the gas.

$$d = \frac{\text{solubility } S}{\sqrt{\text{molecular weight}}}$$

$$V_{gas} = \frac{d \times \text{area} \times (P_1 - P_2)}{\text{thickness}}$$

Diffusion rates of oxygen and carbon dioxide

CO_2 diffuses about 20% slower than O_2 in the gas phase but as it is about 24 times more soluble it diffuses 20 times faster across the alveolar membrane.

Pulmonary diffusing capacity

Because the area and thickness over which diffusion occurs is unknown, a gas transfer index (DL) is used to overcome this and is the volume (V) of gas taken up per unit time divided by the pressure gradient for the gas across the alveolar–capillary membranes.

$$DL = V/(P_1 - P_2)$$

Thus DL takes into account the area and the thickness of the diffusion path, but also the diffusing characteristics of the gas.

For oxygen this would be:

$$DL_{O_2} = V_{O_2}/(P_{A_{O_2}} - P_{C_{O_2}})$$

The mean capillary P_{O_2} (averaged over the whole length of the capillary) is difficult to measure because of its not inconsiderable solubility in plasma and the kinetics and binding with haemoglobin. Instead, carbon monoxide (CO) is used because of the following properties:

- CO follows the same route as O_2. Its partial pressure in the pulmonary capillary is so small it can be neglected (except in smokers).
- Its affinity for haemoglobin is more than 200 times that of O_2, which means that for small quantities of CO the latter will be preferentially bound to haemoglobin and so the P_{co} of capillary blood is effectively zero.
- Diffusing capacity for CO = CO uptake/alveolar CO.

Measurement of pulmonary diffusion capacity (DL_{co})
There are two methods of measuring pulmonary diffusion capacity:
1. Single breath method.

2. Steady-state method:
   a. subject breathes 0.1% CO for about 0.5 min;
   b. constant rate of disappearance of CO from alveolar gas measured;
   c. alveolar concentration of CO measured at end of each expiration.

Factors affecting DL_{co}
Several factors affect the pulmonary diffusion capacity:

- Haemoglobin concentration, e.g. DL_{co} is low in anaemia (fewer Hb binding sites).
- Pulmonary blood volume – increase in pulmonary volume, e.g. exercise or blood shift into the thorax, increases DL_{co} (more Hb available to take up CO).
- Larger pulmonary surface area increases DL_{co} (related to body size).
- Prevailing Po_2 in alveolar air because of competition between CO and O_2 for haemoglobin, the higher the Po_2 the lower is DL_{co}.

$$DL_{co} = V_{co} \text{ (ml/min)}/PA_{co} \text{ (mmHg)}$$
$$DLo_2/DL_{co} = So_2/S_{co} \times \sqrt{MW_{co}}/\sqrt{MWo_2}$$
$$= 0.024/0.018 \times 28/32$$
$$= 1.23$$
$$DLo_2 \quad = 1.23 \times DL_{co}$$

TRANSPORT OF OXYGEN AND CARBON DIOXIDE IN THE BLOOD

Partial pressures of gases in alveoli and blood

	mmHg O_2	kPa	mmHg CO_2	kPa	mmHg N_2	kPa	mmHg H_2O	kPa
Alveolar air	100	13.3	40	5.3	573	76.4	47	6.3
End pulmonary capillary	100	13.3	40	5.3	573	76.4	—	—
Arterial blood	95	12.7	40	5.3	573	76.4	—	—
Mixed venous blood	40	5.3	46	6.1	573	76.4	—	—

Content of gases in arterial and mixed venous blood

	O_2 (cm³/100 ml)	CO_2 (cm³/100 ml)
Arterial blood	19	54
Mixed venous blood	14	58

CARRIAGE OF OXYGEN IN THE BLOOD

Oxygen is carried in the blood in physical solution and combined with haemoglobin (Hb).

In physical solution

$$O_2 \text{ (dissolved)} = \alpha Po_2$$

α = solubility constant
= 0.023 ml O_2/ml at 760 mmHg and 37°C
O_2 dissolved in blood plasma when:
alveolar O_2 = 14.5%: barometric pressure = 760 mmHg
saturated water vapour pressure at 37°C = 47 mmHg
Po_2 = % $O_2 \times$ gas pressure

$$= \frac{14.5}{100} \times (760 - 47) \text{ mmHg} = 103.4 \text{ mmHg}$$

Volume of O_2 dissolved in 1 ml blood:

$$= \frac{0.023 \times 103.4}{(760 - 47)} = 0.003 \text{ ml}$$

Combined with haemoglobin

1 mol/L (32 g) O_2 combines with 16 700 g Hb
but 1 mol/L O_2 occupies 22.4 L at STP

1 g Hb combines with $\dfrac{22\,400 \text{ ml/mol}}{16\,700 \text{ g per equiv. Hb}}$ = 1.34 mol O_2

Alternatively:
100 ml blood (15 g Hb) combines with 20.1 ml O_2

1 g Hb combines with $\dfrac{20.1}{15}$ = 1.34 ml O_2

The quantity of O_2 carried in physical solution is not capable of supplying the O_2 required for the body's metabolic needs (250 ml O_2 at rest). A cardiac output of some 80 L/min or more would be required to supply metabolic needs if physical solution was the only method of transporting O_2.

Haemoglobin (Hb)
It is the Hb inside the red cell which allows blood to carry so much oxygen. Hb combines with O_2 in a loose chemical combination which is an *oxygenation* and not an oxidation.

Hb is made up of four iron-containing haem groups each combined with a globulin. Each globulin molecule is made up of four polypeptide chains. In normal adults, Hb (HbA) consists of:
- Two α-chains (each of 141 amino acids).
- Two β-chains (each of 146 amino acids).

The amino acid chains making up the globulin form a complex three-dimensional structure and in each chain is a crevice which contains a haem group. *Haem* consists of a *protoporphyrin ring* surrounding an iron atom in its ferrous state (Fe^{2+}). The iron is bonded to four nitrogen atoms of the porphyrin ring, to the proximal histidine of a polypeptide chain and loosely and reversibly to O_2. The globin can bind CO_2 and buffer H^+ and the β-chains bind to *2,3-biphosphoglycerate (2,3-BPG)* [2,3-diphosphoglycerate (2,3-DPG)].

When combined with oxygen, Hb is said to be oxygenated and is in the *relaxed state*. When Hb gives up its oxygen it is said to be deoxygenated and is in the *tense state*. In the deoxy (tense or T) state strong ionic bonds form between the polypeptide chains, making them immobile and keeping them apart. This has the consequence of pushing the iron atom out of the plane of the haem ring. Once an oxygen atom binds to a haem the electronic state of the iron atom changes, its diameter increases and it moves (0.08 nm) into the plane of the haem ring. This causes movement of the proximal histidine and the α-helix which contains it; this conformation change is transmitted to neighbouring subunits (as ionic bonds between polypeptide chains become broken), so the whole Hb molecule undergoes a transition from deoxy (T) to oxy (R) states (an allosteric interaction). Therefore the attachment of one O_2 molecule facilitates the uptake of subsequent ones. This cooperative effect on the affinity of O_2 for haemoglobin is called the *haem–haem interaction* and is the reason for the sigmoid shape of the oxyhaemoglobin dissociation curve.

The oxyhaemoglobin dissociation curve (Fig. 3.2)
This curve is the graphical representation of the percentage saturation of haemoglobin and the partial pressure of oxygen, or the oxygen content plotted against partial pressure.

Some related definitions are:

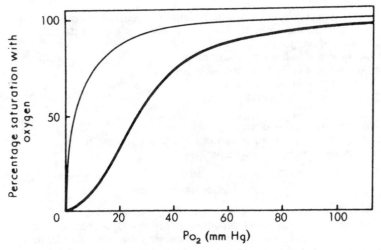

Fig. 3.2
Oxygen-haemoglobin dissociation curve.

Oxygen capacity   The O_2 capacity is the maximal amount of O_2 that will combine with Hb at a high P_{O_2}.
Thus the capacity will vary with the Hb concentration (see above).
Oxygen saturation   The O_2 saturation is the percentage of the O_2 capacity that is occupied by O_2:

$$O_2 \text{ saturation} = \frac{O_2 \text{ bound to Hb}}{O_2 \text{ capacity}} \times 100$$

Note that saturation does not depend upon the concentration of Hb.
Oxygen content O_2 content is the sum of the O_2 combined with Hb and that in physical solution.

Combination of oxygen with haemoglobin
Each haem group is a binding site for O_2.
The oxygenation proceeds in four separate stages.
The affinity for O_2 increases with each stage.

Note the following   The steepest part of the oxyhaemoglobin dissociation curve lies between 10 and 50 mmHg. The slope is shallow above 70 mmHg, so that even moderate reductions in arterial P_{O_2} do not reduce significantly the amount of O_2 carried, e.g. playing wind instruments, breath holding and ascent to altitude. As the P_{O_2} falls when blood reaches the tissues, large quantities of O_2 are unloaded. On arrival of mixed venous blood at the lung ($P_{O_2} \cong 6.1$ kPa or 46 mmHg) oxygen uptake occurs.

Factors facilitating the release and uptake of oxygen by haemoglobin
Factors which move the oxyhaemoglobin dissociation curve to the right increase the amount of oxygen evolved at a given partial pressure of oxygen:
• An increase in the P_{CO_2} (Bohr effect).
• An increase in $[H^+]$ – decrease in pH.
• An increase in temperature.
• 2,3-Biphosphoglycerate.

The role of 2,3-biphosphoglycerate
2,3-Biphosphoglycerate is synthesised in the red cell from metabolites of the glycolytic pathway:

1,3-Biphosphoglycerate $\rightarrow$ 2,3-Biphosphoglycerate
                   Biphosphoglyceromutase

It binds tightly to deoxy (T) Hb. In the presence of biphosphoglyceromutase the affinity of Hb for O_2 is reduced. The dissociation curve shifts to the right causing the release of O_2 into the tissues. Synthesis is stimulated by hypoxia (anaemia and altitude hypoxia).

Partition of oxygen in blood (at rest)

	Arterial blood	Mixed venous blood
O_2 content (ml/100 ml) whole blood	19	14
Partial pressure (mmHg)	95	40
% saturation of Hb	95	70
Dissolved in plasma (ml/100 ml blood)	0.3	0.12

The oxygen cascade
Oxygen moves down a partial pressure gradient from air to mitochondria.

Air→respiratory tract→alveoli→arterial blood→systemic capillaries→the cell→ mitochondria

THE CARRIAGE OF CARBON DIOXIDE IN THE BLOOD

Carbon dioxide is transported in the following forms:
- Dissolved CO_2.
- Carbonic acid.
- Bicarbonate ions.
- Carbamino compounds.
  CO_2 diffuses out of the tissue cells into plasma where it is:
- Physically dissolved.
- Converted to carbonic acid.
- Converted to carbamino compounds.
- Enters the red cells.

Physically dissolved = α x Pco$_2$
α is the solubility coefficient = 0.03 mM/L/mmHg.
Dissolved in arterial blood = $0.03 \times 40 = 1.2$ mM/L.
Dissolved in mixed venous blood = $0.03 \times 46 = 1.38$ mM/L.
Dissolved CO_2 accounts for some 10% of total CO_2 exchange in the lung.

Converted to carbonic acid

$$CO_2 + H_2O \Leftrightarrow H_2CO_3$$

A negligible amount of carbon dioxide is transported in this form.

Entry into the red cell
CO_2 reacts with water to form carbonic acid. This is a very slow reaction in plasma, but in the red cell it is speeded up by the enzyme carbonic anhydrase.

$$CO_2 + H_2O \Leftrightarrow H_2CO_3 \Leftrightarrow H^+ + HCO_3^-$$
$$\text{carbonic anhydrase}$$

This reaction proceeds to the right because the products of the reaction are removed:

1. H^+ is removed by the buffering action of haemoglobin. When Hb loses its oxygen it becomes more basic

$$KHbO_2 + H^+ \rightarrow K^+ + HHb + O_2$$

2. $[HCO_3^-]$ in the red cell increases above that of plasma and so HCO_3^- diffuses out of the cell and is replaced by Cl^- from plasma to maintain electrical neutrality of the red cell (chloride shift – Hamburger phenomenon). About 90% of the CO_2 transported by blood is carried as bicarbonate which has a concentration of 24 mM/L of plasma and 21.5 mM/L as whole blood.

Carbamino transport

About 5% of the carbon dioxide is carried as carbamino compounds, very little is formed in the plasma, the arterial and venous plasma have the same small concentration.

In the red cell CO_2 combines with the amino groups of Hb:

$$CO_2 + HbNH_2 \Leftrightarrow HbNHCOOH \Leftrightarrow HbNHCOO^- + H^+$$

HHb combines with more CO_2 than HbO_2, so unloading occurs rapidly in the lungs and is facilitated in the peripheral tissues CO_2.

	Dissolved CO_2	Carbamino CO_2	Bicarbonate
Proportions of CO_2 unloaded by the blood	10%	30%	60%

Note:
1. The CO_2 dissociation curve is more linear than the O_2 dissociation curve.
2. Deoxygenated blood carries more CO_2 than oxygenated blood for a given P_{CO_2}, due to the better ability of reduced Hb to mop up more H^+:

Figure 3.3 is a diagrammatic summary of the role of the red cell in O_2 and CO_2 transport.

THE PULMONARY CIRCULATION

The function of the pulmonary circulation is to convey blood to the lungs through the pulmonary artery and back to the heart by the pulmonary veins, so that a rapid and efficient exchange of gases can occur across the alveolar epithelium. It differs from many of the circulations through other organs:

1. Almost the entire output of the right ventricle goes to the alveolar capillaries.
2. The density of the capillary network presents a huge surface area for gas exchange, about 90–125 m^2 in the adult.
3. There is a very low vascular resistance with less smooth muscle in the arterioles and small arteries and therefore they only play a minor role in regulation of blood flow. Resistance is about one-eighth of that of the systemic circulation.

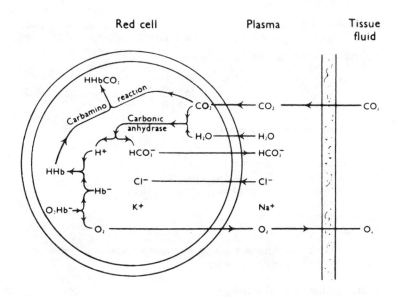

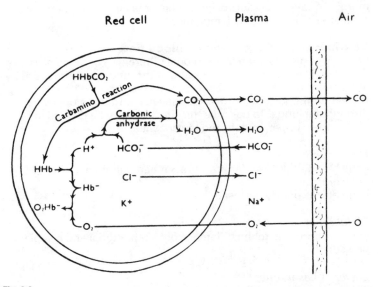

Fig. 3.3
Summary of the role of the red cell in oxygen and carbon dioxide transport.
(Reproduced with permission from: Roughton FJW 1935 Physiological Reviews 15: 293.)

4. No autoregulation.

$$\text{Pulmonary vascular resistance} = \frac{\text{Mean arterial pressure} - \text{Left atrial pressure}}{\text{Blood flow}}$$

$$= \frac{12-5 \text{ mmHg}}{5 \text{ L/min}} = 1.4 \text{ mmHg/L/min}$$

Pressures in pulmonary vessels

Pulmonary artery
Mean pressure 11–15 mmHg.
Systolic pressure 20–23 mmHg.
Diastolic pressure 5–9 mmHg.

Lung capillaries
5–9 mmHg

As the cardiac output rises, pressure changes little due to the recruitment of empty capillaries.

Water balance in the lungs

At rest the maximum capillary pressure is less than the colloid osmotic pressure of the plasma proteins (15 mmHg versus 25 mmHg), therefore there is no net filtration. In exercise this margin of safety is reduced. Fluid which may leak from the capillaries is removed by lymphatics.

Distribution of the pulmonary blood flow

Because of gravity, there are regional differences in blood pressure and flow. The mean arterial pressure at the lung apex is reduced by about 11 mmHg and at the base increased by about 11 mmHg. Thus in the upright position there is an uneven distribution of flow through the lungs. To describe this the lung may be divided into three zones, upper, middle and lower zones.

Upper zone
Little blood flow (alveolar pressure > mean pulmonary arterial pressure).
Lung apices poorly perfused.

Middle zone
Flow proportional to the difference between arterial and alveolar pressures.

Lower or basal zone
Flow proportional to the difference between pulmonary arterial and pulmonary venous pressures.

Regional differences in flow become less during exercise.

Flow in pulmonary vessels

Output of right ventricle:
  45% to left lung
  55% to right lung.

Circulation time

Pulmonary artery to left atrium:
  5 s.
Time traversing a pulmonary capillary:
  at rest 0.75 s.
  during exercise 0.3 s.

Pulmonary blood volume

0.5–1L or about 10–20% of total blood volume.
100 ml in pulmonary capillaries.
Volume very variable.
Transfer of blood from pulmonary to systemic vessels varies:
  Output from left to right ventricle can be momentarily unequal –
  lungs act as temporary reservoir
  Less in the pulmonary circulation when erect compared with
  supine.
Increases in pulmonary volume:
  Increases in left atrial pressure – mitral stenosis
  Left ventricular failure.

Changes with breathing

Inspiration
Fall in intrathoracic pressure→increased right ventricular filling →
increased ventricular output (Starling effect)→output passes into an
increase in capacity of pulmonary bed (intrapleural vessels
distend)→output from right side not immediately transmitted to the
left side of the heart and stroke volume may transiently fall→small
fall in blood pressure.
  The actual result depends upon a balance between the increase
in the venous return and the enlarged capacity of the pulmonary
circulation on inspiration.

Expiration
During expiration the above changes are reversed.
  The mismatch between venous return and the transmission to
the left ventricle is the cause of the split second sound of the heart.

Control of pulmonary circulation

The control of vascular resistance and volume flows are determined
largely by passive factors: neither nerves nor metabolic mediators
seem to affect pulmonary blood flow under normal conditions.

Hypoxia
When the Po_2 of the alveoli falls, arterioles close to, and supplying

these alveoli vasoconstrict, which diverts blood away from them and to better ventilated areas. This then equalises regional ventilation/perfusion ratios in the lung, constriction to hypoxia is abolished when endothelium is removed from isolated arteries.

The mediator does not appear to be nitric oxide, or endothelin or a prostanoid. Adenosine may be implicated as its effect is reduced by an adenosine antagonist in the rat.

Chronic hypoxia
Residence at high altitude→chronic hypoxia→pulmonary hypertension
- Breathing is deeper.
- Reduction in intrathoracic pressure may lead to:
    pulmonary oedema
    right ventricular failure.

BRONCHIAL CIRCULATION

The trachea, bronchi and bronchioles are supplied by the bronchial arteries ($\cong$1% of cardiac output). Venous drainage is into:
- Bronchial veins (25%).
- Pulmonary veins (75%).

There are some direct connections between bronchial, myocardial (thebesian veins) and pulmonary circulations (downstream of the pulmonary capillaries), which account for the small degree of desaturation of the pulmonary venous blood. This admixture is normal and is called a *right-to-left shunt*.

Ventilation and perfusion
Ventilation and perfusion increase from apex to base, due to the effect of gravity. The effect on perfusion is greater because blood has a larger specific gravity. The degree of matching of these two variables is expressed as the ventilation/perfusion ratio ($\dot{V}_A/\dot{Q}$). Typical values in a standing individual:
- Base of lung 0.5.
- Apex of lung 3.0.

The average $\dot{V}_A/\dot{Q}$ = alveolar ventilation ($\dot{V}_A$)/cardiac output ($\dot{Q}$)
= 4 L per min/5 L per min = 0.8

$\dot{V}_A/\dot{Q}$ ratios vary from infinity (∞), dead space no perfusion, to zero (no air entering well-perfused alveoli).

Non-respiratory and metabolic functions of the lung
In addition to gas exchange, the lung has other functions:
1. The conversion of angiotensin I→angiotensin II converting enzyme.
2. Prostaglandins, 5-hydroxytryptamine, bradykinin and leukotrienes are removed.
3. Noradrenaline is 30% removed in a single pass but not adrenaline.

4. Production of nitric oxide.
5. Synthesis of phospholipid – dipalmitoyl lecithin – component of surfactant.
6. Protein synthesis – collagen and elastin.
7. Elaboration of mucopolysaccharides – bronchial mucus.
8. Secretion of immunoglobulins in bronchial mucus.

THE CONTROL OF VENTILATION

The breathing rhythm begins at birth and continues until death. How this rhythm is generated and maintained has even now not yet been satisfactorily explained. Gas exchange is accomplished by the rhythmical movements of intercostal muscles, diaphragm and abdominal muscles enlarging the chest cavity and which are brought about by nerves. These neurones innervate the:
- Diaphragm (the phrenic nerve) at (C_3–C_6).
- Intercostal muscles (T_1–T_{12}).
- Abdominal muscles (T_4–L_3).
  The motor neurones controlling airway dimensions are in the brain stem:
- Alae nasae – the facial nucleus.
- Pharyngeal muscles – the glossopharyngeal and the glossopharyngeal branch of the vagus.
- Bronchial smooth muscle – from the vagus.
- The larynx – the recurrent laryngeal nerve.
  2, 3 and 4 come from the nucleus ambiguus.
  These nerves also have non-respiratory functions of:
- Phonation.
- Protective reflexes of the upper airways (coughing, sneezing, swallowing).
- Involved in other responses (vomiting, defecation, parturition and micturition).

LOCATION OF CENTRAL NEURONES CONTROLLING VENTILATION (Fig. 3.4)

A rhythm generator is located within the medulla, which appears to be a network which requires at least six neuronal populations. Breathing is normally automatic, but it is also under voluntary control (speech, singing), and therefore involves the cerebral cortex.

Respiratory neurones, which discharge in synchrony with the cycle of inspiration and expiration, form two parallel columns in the medulla, these are:
1. A dorsal respiratory group (DRG), part of the nucleus of the tractus solitarius, which is mainly inspiratory and projects to the

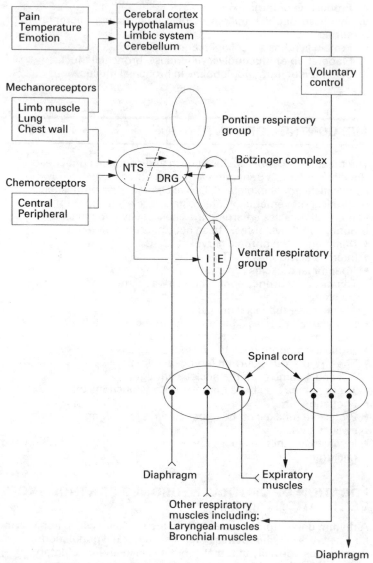

Fig. 3.4
The major neurone groups controlling ventilation with possible connections between these and to peripheral and central receptors. DRG = Dorsal respiratory group; NTS = nucleus tractus solitarius. (Adapted from: Bray et al 1994 Lecture notes on human physiology. Blackwell Science, Oxford.)

spinal cord to synapse with neurones in the phrenic nerve (C_3–C_6). The DRG receives an input from both mechanoreceptors (chest wall and lung) and chemoreceptors.

2. A ventral respiratory group (VRG) located in the nucleus ambiguus (NA) and nucleus retroambigualis (NRA). The NA and the rostral part of the NRA contain inspiratory neurones and mainly supply external intercostals and accessory muscles of respiration. These neurones receive a sensory input and behave like those of the DRG.

The caudal part of the NRA area contains neurones which fire in synchrony with expiration, they fire in the quiescent period between bursts of phrenic action potentials. Neurones from this part project (premotor neurones) to supply internal intercostal and abdominal muscles.

There are at least two other groups of nuclei in the brain stem involved in ventilatory control. These are a pontine respiratory group (PRG) formerly known as the pontine centre, and a group of neurones known as the Bötzinger complex. The PRG neurones do not seem essential for respiratory rhythm generation, their function is to:

- Stabilise the respiratory pattern.
- Slow the rhythm and influence the timing of the respiratory phases.
- Act as a relay nucleus to convey information from the hypothalamus to the respiratory centres.

THE BÖTZINGER COMPLEX

It is believed that in and around this complex are the neurones responsible for the generation of the breathing rhythm, because injection into this area of a neurotoxin which does not destroy passing axons causes the respiratory rhythm to stop.

HIGHER CENTRES IN THE BRAIN INFLUENCING VENTILATION

Hypothalamus
Provides a continual background drive to the DRG.
Central and peripheral thermoreceptors – influence breathing, e.g. fever increases breathing.
Activating the *defence reacting centre*→increases breathing.
Descending pathways from hypothalamus end in the PRG.

Cerebral cortex
Provides an inhibitory drive to the hypothalamus and the DRG.
Emotional and painful stimuli which affect respiration arise in the cortex and limbic system to impinge on the hypothalamus.
Voluntary control of respiration, phonation, breathholding, and breath control for playing wind instruments stems from the cortex,

with the neural connections by-passing the medullary control mechanisms. These are short-term events and reflex control eventually takes over.

CONTROL OF RATE AND DEPTH OF VENTILATION

Movements of the chest wall and diaphragm which determine the depth and rate at which the lungs are ventilated depend on two major factors:
1. The arterial concentration of CO_2 (alveolar CO_2).
2. Oxygen.

Note: Alveolar air is in equilibrium with arterial blood. End tidal air is often used to measure arterial gas concentrations.

Ventilation is modified by information arising from:
- Central inputs.
- Cerebral cortex – see above.
- Limbic system and hypothalamus – see above.

CENTRAL CHEMOSENSITIVE REGIONS

These are situated on the ventrolateral surface of the medulla, bounded by the pontine–medullary junction, the roots of the VIIth to XIth cranial nerves and the pyramids. The receptors (not yet identified) appear to lie about 500 µm from the surface of the medulla.

These receptors can be influenced from two sources:
1. The blood through the interstitial fluid (ISF), but separated by the blood–brain barrier which is relatively impermeable to charged molecules and ions, e.g. H^+ and HCO_3^-.
2. The CSF bathing the surface of the medulla through the ISF by pH and P_{CO_2}.

These receptors are stimulated by arterial hypercapnia, which is responsible for 80% of the ventilation due to hypercapnia, but *are insensitive to hypoxia*.

As the CSF is some distance from the chemoreceptors its influence upon them is slower than changes in arterial blood and must be taken into account when interpreting the ventilation changes due to acid–base disturbances.

PERIPHERAL INPUTS

Receptors in the periphery input to the respiratory areas. These originate from peripheral chemoreceptors and peripheral mechanoreceptors.

Peripheral chemoreceptors
These are the carotid body and aortic bodies (Figs 3.5, 3.6), and are stimulated by arterial:
- Hypoxia.

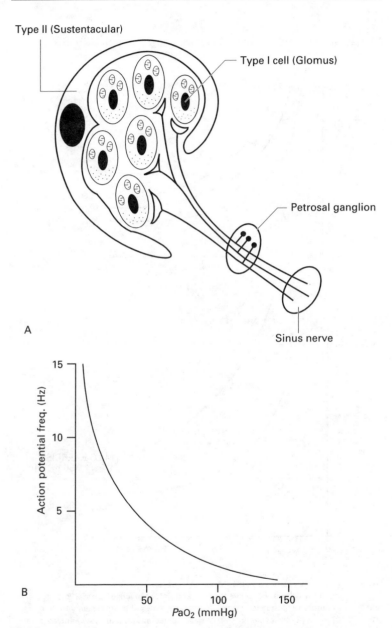

Fig. 3.5
The carotid body (A) and the discharge from the sinus nerve (B). Note that the nerve discharges at a rate proportional to the degree of hypoxia. (Adapted from: Biscoe TJ, Duchen MR 1990 News in Physiological Science 5: 229–233.)

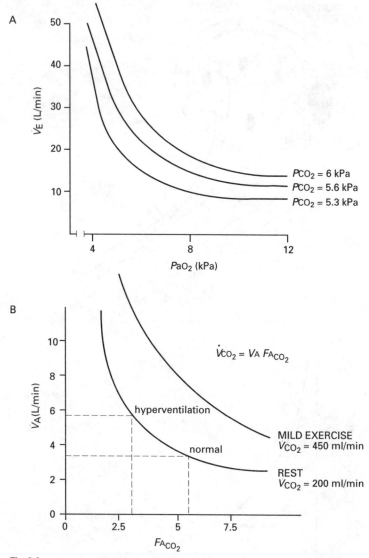

Fig. 3.6
A, The effect of the arterial Po_2 on ventilation at different levels of arterial Pco_2. Note that the ventilatory response to hypoxia becomes greater at progressively higher arterial Pco_2. B, Hyperbolic relationship between alveolar ventilation ($\dot{V}A$) and alveolar CO_2 concentration (FA_{CO_2}), which defines the rate of CO_2 production ($\dot{V}co_2$). Thus a given $\dot{V}co_2$ can be attained by an infinite combination of VA and FA_{CO_2}. Increased $\dot{V}co_2$, such as seen during muscle exercise, shifts the Vco_2 isopleth upwards, as shown. (B, reproduced with permission from Whipp: The Respiratory System)

- Hypercapnia.
- Acidity.
- Their contribution to ventilation:
   the total response to acute hypoxia
   about 20% of the response to acute hypercapnia
   nearly all the response to acute metabolic acidaemia.

The carotid bodies
These are situated at the bifurcation of the internal and external carotid arteries. Each has a dense capillary network with a blood flow of 20 ml/g/min. Surrounding the capillaries are *glomus cells* and *sustentacular cells* (supporting cells) and sensory nerve endings.

The aortic bodies
These are numerous over the arch of the aorta and main arteries in that area. The structure is similar to the carotid bodies, though they are less vascular. Cardiovascular responses are more prominent than respiratory responses. These receptors are not as sensitive to hypoxia, hypercapnia or acidity as the carotid bodies.

Peripheral mechanoreceptors

Lung reflexes
These reflexes are initiated by stretch receptors. Lung inflation leads to inhibition of inspiratory activity (Hering–Breuer reflex), which is abolished by vagal section in the neck to give rise to slower and deeper breathing. Pulmonary stretch receptors are located in the bronchi to alveolar ducts, and are slowly adapting. This inflation reflex is weak in humans.

Chest wall proprioceptors
These reflexes also influence the breathing pattern and provide information about lung inflation. The receptors are situated in joints, tendons (Golgi) and muscle spindles located in the muscles and joints in the chest wall and diaphragm. They provide feedback so that the strength of contraction can be varied to overcome airway resistance changes. They may also be involved in the sensation of dyspnoea (difficulty in breathing) and therefore also project to the cerebral cortex.

Irritant receptors
Rapidly adapting receptors situated in the epithelium are attached to myelinated vagal afferent fibres. They respond to chemical irritants, dust, noxious gases and tobacco smoke, but can also be stimulated mechanically. The response is hyperpnoea, bronchoconstriction and coughing. It has been suggested that in asthma the histamine which is released stimulates these receptors.

C-receptors (pulmonary and bronchial) or J-receptors (juxtacapillary)
These receptors lie in the interstitium or next to capillaries and are attached to unmyelinated C-fibres. These do not play a role in the normal control of breathing but are stimulated by interstitial distortion or congestion. The reflex causes bronchoconstriction and apnoea, followed by rapid shallow breathing.

Limb proprioceptors
Impulses from muscle spindles and other limb proprioceptors have been implicated in the hyperpnoea of muscular exercise.

ARTIFICIAL VENTILATION

This is required when breathing is absent or severely depressed, e.g. in:
• Anaesthesia.
• Cardiac arrest.
• Head injury.
• After administration of a muscle relaxant.
• Acute respiratory failure associated with critical illness.

Mechanical ventilation
Mechanical ventilation is a routine clinical procedure. It requires:
• Intermittent positive pressure (IPP) through a cuffed endotracheal tube.
• Continuous human supervision.
• Alarms to detect low and high pressure limits in the system.
• Pulse oxymetry to measure saturation of haemoglobin.
• Measurement of end tidal CO_2 – capnography.

Artificial ventilation in emergencies
Artificial ventilation is normally required following:
• Drowning.
• Cardiac arrest.
• Electric shock.
• Carbon monoxide and drug poisoning.

Principles
The heart will continue to beat independently of the central nervous system but *prompt action is required*. Arterial O_2 content falls from 20 ml/dl to 15 ml/dl in 60 s and to zero within 5 min. Cerebral function is rarely restored after 2–3 min (except under conditions of hypothermia). In cardiac arrest (drowning, electric shock, coronary occlusion, etc.) external cardiac massage and artificial ventilation is required. During these procedures it is essential to:
• Ensure an open airway.
• Breathe for casualty.
• Circulate the blood – by intermittent compression of the chest.

Methods for expired air resuscitation ('mouth to mouth')
The following steps are involved in mouth-to-mouth resuscitation:
Casualty placed on back on firm surface.
Mucus and any debris removed from mouth and throat.
Head extended as far as possible to open up airway.
Casualty's nose pinched to prevent leakage of air.
Operator blows into casualty's mouth until chest moves.
Operator takes mouth away quickly and allows passive deflation of casualty's lungs.
Process repeated 10 times/min.
After first four ventilations see if the heart is beating.

Checking for heartbeat
Casualty may be blue around lips.
Place finger between muscles and larynx – can the carotid pulse be felt?
Check for pulse every 3 min thereafter.

External chest compression
Place the heel of one hand on the centre of the lower half of the sternum (breast bone).
Cover this with the heel of the other hand and lock the fingers together.
Press by rocking forwards and backwards with the arms straight – a movement of the chest of about 1.5–2 inches (4–5 cm) will occur in the adult. Carry out six to eight compressions to one mouth inflation.
When the pulse returns carry on with mouth-to-mouth ventilation only until breathing is spontaneous.
If no response continue both manoeuvres until expert help arrives.

HYPOXIA

This is a term given to denote that the partial pressure of oxygen in the arterial blood is below the normal range of about 90–100 mmHg (perhaps the proper term should be hypoxaemia).

Hypoxic hypoxia
O_2 content of blood reduced because Pa_{O_2} is reduced.

Causes:
Inequality of $\dot{V}a/\dot{Q}$.
Obstruction in air passages.
Paralysis of respiratory muscles.
Breathing gases of low O_2 content (e.g. at high altitude).
Alveolar–capillary block.
Reduction in area available for gas diffusion.

Anaemic hypoxia
Interference with O_2 transport.

Causes:
Reduction of circulating haemoglobin (anaemia).
Some haemoglobin unavailable for gas transport (e.g. methaemoglobin).

Compensatory mechanisms
Increase in cardiac output.
Increase in circulation rate.
Shift of O_2 dissociation curve to the right (an increase in 2,3-BPG in anaemia).
Patient breathless in mild exercise, but not at rest.

Stagnant hypoxia
Local slowing of the circulation.

Causes:
Vasoconstriction – too cold.
Cardiac failure.
Peripheral circulatory failure.

Histoxic hypoxia
Interference with O_2 utilisation by the tissues, e.g. cyanide poisoning.

Cyanosis
A violet, bluish or greyish colour of the skin consequent upon abnormally low reduced Hb (5 g reduced Hb/100 dl blood).

Central cyanosis
Inadequate oxygenation, decreased P_{O_2} of inspired gas.
Hypoventilation.
Lung disease.
Right-to-left heart shunt.
Extremities warm and pulsatile.
Peripheral blood flow rapid.
Tachycardia.
Pulse pressure increased.

Peripheral cyanosis
Extremities cold and blue.
Peripheral pulses difficult to detect.
Blood flow reduced at the periphery (vasoconstriction).
Cyanosis due to long residence time (stagnant hypoxia).

Acute hypoxia
The majority of persons who ascend rapidly to terrestrial elevations

higher than 2500 m (8200 feet) undergo an unpleasant experience known as mountain sickness, the most prominent symptoms of which are: headache, nausea and vomiting, and insomnia.

Usually after a period of 3–7 days a process of acclimatisation occurs and the symptoms abate.

At this altitude, but usually at those much greater, a proportion of climbers will experience life-threatening diseases resulting in pulmonary oedema and cerebral oedema.

Hypoxia stimulates ventilation when alveolar Po_2 falls below 50–60 mmHg.

The reflex drive from peripheral chemoreceptors is illustrated below.

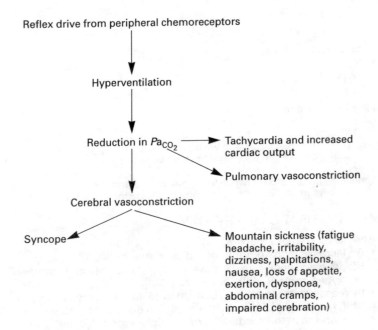

ASCENT TO ALTITUDE

During a slow ascent to altitude (3700 m, 12000 ft) ventilation is increased. This is the result of a balance between the hypoxic drive and drive from peripheral chemoreceptors and the pH drive from central chemoreceptors.

An example of a slow ascent to altitude is given below.

The mountaineer
A slow ascent to 12 000 feet (3700 m)

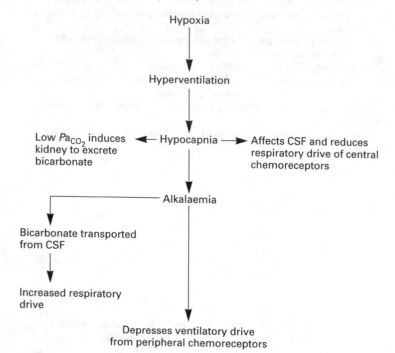

Ventilation is a balance between the hypoxic drive from peripheral chemoreceptors and the 'pH drive' from central chemoreceptors

ACCLIMATISATION TO HYPOXIA

After several weeks there is:
- An increase in the number of red cells and Hb (8×10^6/L: 21 g Hb/dl) which allows an increase in oxygen capacity and increase in oxygen content of the blood (28 ml/dl).
- Increased erythropoiesis in bone marrow (through erythropoietin).
- Shift to the right of the oxyhaemoglobin dissociation curve.
- An increase in the production of 2,3-biglycerophosphate.
- An increase in the number of capillaries.
- Redistribution of blood to more vital tissues.
  Some consequences of chronic hypoxia are shown below.

PROPHYLAXIS

Most of the symptoms of mountain sickness can be prevented by taking the carbonic anhydrase inhibitor acetazolamide, before ascending to altitude, but the best advice is not to go 'too high too fast'.

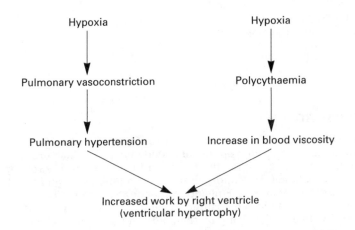

THE CEREBROSPINAL FLUID (CSF)

The brain is covered by the pia mater. The pia and the arachnoid are separated by a space through which pass strands of arachnoid connecting the two, and which contains a fluid the CSF. The subarachnoid space is continuous with the ventricular system through the medial and lateral apertures in the roof of the fourth ventricle. At the base of the brain, pia and arachnoid are widely separated to form cisterns, the largest of which is the *cisterna magna*. Continuous with the fourth ventricle is the central canal of the spinal cord. The subarachnoid space extends beyond the end of the spinal cord containing the *cauda equina*. It is at this site between the third and fourth lumbar vertebrae that a needle may be passed to obtain a sample of CSF (lumbar puncture) without damaging the cord.

COMPOSITION OF CSF

Composition	Blood plasma	CSF
Protein	60–80 g/L	200–400 mg/L
Glucose (fasting)	3.0–5.0 mmol/L	2.5–4.5 mmol/L
Urea	2.5–6.5 mmol/L	2.0–7.0 mmol/L
Sodium	136–148 mmol/L	144–152 mmol/L
Potassium	3.5–5.0 mmol/L	2.0–3.0 mmol/L
Calcium	2.2–2.6 mmol/L	1.1–1.3 mmol/L
Chloride	95–105 mmol/L	123–128 mmol/L
Bicarbonate	24–32 mmol/L	24–32 mmol/L

Note: The composition is similar to that of blood plasma, except for protein and chloride. If calcium is expressed as ionised calcium, the concentration in blood and CSF are similar.

The CSF is:
- Clear and colourless.
- SG = 1.005.
- pH = 7.33.
- Contains not more than 5 lymphocytes/ml.
- PD of +5 mV between blood and CSF.

FORMATION

The CSF is formed by the choroid plexus. This is a network of capillaries covered with epithelial ependymal cells which project into the ventricles (lateral, third and fourth ventricles). The rate of formation of the CSF is approximately 0.5 ml/min. (720 ml/24 h). It has a total volume of 120 ml and exchanges every 4 h. The pressure of the fluid is measured by lumbar puncture and is 10 cmH_2O with a subject lying on their side and 30 cmH_2O when standing.

Mechanism of formation

The CSF is formed by active transport of Na^+ across the apical membrane utilising the $Na^+ - K^+$ ATPase. Na^+ enters over the basolateral membrane through Na^+/H^+ and Na^+ClCl^- exchangers. Bicarbonate is synthesised in the epithelial cell and is secreted into CSF:

$$CO_2 + H_2O \Leftrightarrow H_2CO_3 \Leftrightarrow H^+ + HCO_3^-$$
(catalysed by carbonic anhydrase)

H^+ then passes out into the blood over the basal membrane in exchange for Na^+.

RE-ABSORPTION OF CSF

The main route for re-absorption is via the venous system (superior sagittal sinus) through the arachnoid villi. The arachnoid villi are:
- Invaginations of the arachnoid into the venous sinuses.
- Labyrinth of cells separate CSF from blood.
- Cells contain vacuoles opening into CSF.
- When pressure in subarachnoid space > that in superior sagittal sinus, vacuoles open into the blood and CSF passes down the pressure gradient into blood.
- When pressure in subarachnoid space < that in sinus, the vacuoles collapse.
  Other routes of absorption:
- From the extensions of the subarachnoid space along spinal nerves into spinal veins.
- Through the ependymal linings of the ventricles.
- Absorption of CSF into the blood is favoured by the colloid osmotic pressure of the the plasma proteins, since CSF contains very little.

BLOOD-BRAIN BARRIER

Not all substances enter the brain with equal facility, particularly ions, and this has give rise to the concept of the blood-brain barrier.

Highly permeable substances
O_2, CO_2, alcohol, barbiturates, glucose and lipophilic substances.

Poorly permeable substances
Ions and most highly dissociated compounds, e.g. amino acids pass very slowly into the brain.

Anatomical basis for the blood-brain barrier
Tight junctions between the endothelial cells of the capillaries of the CNS restrict diffusional exchanges of water-soluble solute.

Structures in which the blood-brain barrier is absent
The circumventricular organs:
- Area postrema.
- Subfornical organ.
- Pineal organ.
- Median eminence.

FUNCTION OF CSF

The CSF functions as a cushion between the soft brain and the rigid cranium. It supports the brain and distributes the force of blows to the head.

There is a reciprocal relationship between the volume of blood in the head and the CSF. Active transport of some substances by the choroid plexus from CSF into blood:
- 5-HT.
- Adrenaline.
- Drugs, e.g. penicillin.

The CSF acts as a sink and solutes at higher concentration in brain diffuse into the CSF then into the blood at the arachnoid villi.

4. The gastrointestinal tract

MOUTH, OESOPHAGUS AND SWALLOWING

THE MOUTH

THE SALIVARY GLANDS

There are three pairs of salivary glands:
* Sublingual.
* Submandibular.
* Parotid.

Composition of saliva
Volume $\cong$ 750 ml/day.
Submandibular $\cong$ 70%.
Parotid $\cong$ 25%.
Sublingual $\cong$ 5%.
pH = 6.2 and is flow dependent – more alkaline at higher rates.
Saliva is hypotonic.

Inorganic constituents
Calcium 1.5 mmol/L.
Phosphorus 5.5 mmol/L.
Na^+, Cl^-, HCO_3^- – flow dependent.
K^+ = 20 mmol/L.

Organic constituents
Amylase, kallikrein, mucin, lysozymes, amino acids, urea, citrate, glycoproteins.
Antigens of ABO, Lewis and SD blood groups, squames, disintegrating white cells, microorganisms.

Function of saliva
Saliva has many functions which include:
* Protection:
    Cleans and protects teeth and oral epithelium
    Washes away debris
    Keeps buccal epithelium moist.
* Amylase action:
    Starch→maltose
    It is likely to have its major action in stomach until pH in food falls.

- Lysozyme activity→reduces bacterial activity.
- Lubrication and moistening of food→facilitates eating, swallowing and speech.
- Dissolves food and facilitates taste.
- Acts as a buffer due to HCO_3^-:
    Prevents fall in pH
    Protects enamel against acid.

Histological structure of the glands

Secretory element of the acinus is a single cell layer around a central cavity. Acini lead into intercalary ducts which in turn form striated ducts (tall cells with basal infoldings). Striated ducts lead into larger excretory ducts→mouth.

Myoepithelial cells

These are specialised cells which embrace the acini and support intercalary ducts. The functions of the myoepithelial cells include:
- Contractile→help to expel saliva.
- Support secretory elements to allow secretion at high pressure (> blood pressure).

Innervation

The salivary glands are innervated by both the parasympathetic and sympathetic nervous systems. The parasympathetic nerves innervate:
- Gland cells.
- Myoepithelial cells.
- Vasodilator to blood vessels.
    The sympathetic nerves innervate:
- Constrictor to blood vessels.

Control of salivary secretion

Secretion of saliva is controlled by:
- Parasympathetic stimulation→profuse salivary secretion and vasodilatation.
- Sympathetic stimulation→viscid secretion, vasoconstriction – role is obscure.

Reflex salivary secretion

Nervous pathway   Salivation can be activated by receptors to external stimuli. Each gland is controlled by a different pathway:
- Sublingual and submandibular glands:

Taste receptor→afferent neurone→superior salivary nuclei→chorda tympani (preganglionic fibre)→submandibular ganglion→ postganglionic fibre→gland cells.

- Parotid gland:

Taste receptor→afferent neurone→inferior salivary nucleus→lesser

superficial petrosal nerve (preganglionic fibre)→otic
ganglion→postganglionic fibre→parotid gland cell.

Reflex salivation
Salivary reflexes are of two types:
1. *Unconditioned reflex*, i.e. present at birth – taste
   a. Stimulation of taste buds together with other stimuli from –
      i. impulses from sensory endings stimulated by mastication
         – teeth
      ii. proprioceptors in masticatory muscles
      iii. oral epithelium;
   b. Distension of the oesophagus;
   c. Stimuli acting from the stomach – gastro-salivary reflex.
2. *Conditioned reflex*, i.e. learned by association with taste:
   a. Sight and smell of food;
   b. Noises associated with the preparation of a meal;
   c. Nervous pathway –
      Cerebral cortex→hypothalamus→salivary nuclei→salivary
glands.

Vasodilatation
Secretion is accompanied by vasodilatation in the salivary glands.
Administering atropine (parasympathetic blockade) abolishes the
secretory response, but vasodilatation is unaffected.
   Two theories are advanced to explain this:

Kinin theory

Protein	$\xrightarrow{\text{kallikrein}}$	kinin
Kininogen		bradykinin

Polypetide neurotransmitter
Vasoactive intestinal polypeptide may be released with
acetylcholine (ACh) from nerve terminals.

Stimulus secretion-coupling
Neurotransmitters (ACh and noradrenaline) react with specific
membrane receptors. This results in an increased membrane
conductance and the influx of Ca^{2+} and Na^+. The rise in Ca^{2+}
initiates the exocytosis of preformed amylase-containing vesicles.
This is accompanied by the breakdown of GTP to cGMP. Activation
of β-receptors has little effect on membrane potential but activates
adenylate cyclase, leading to a rise in cAMP (cyclic AMP). This is
linked with a rise in Ca^{2+} leading to exocytosis and the release of
amylase.

THE OESOPHAGUS

Anatomy
25 cm long by 2 cm in diameter.
Extends from pharynx to cardiac end of stomach.

Lined by stratified squamous epithelium.
The mucosa exhibits longitudinal folds when empty.
Mucous glands exist in the submucosa.
Abdominal segment is lined with gastric-type mucosa.

Muscle
Upper third→striated muscle.
Lower third→smooth muscle.
Middle third→mixture of striated and smooth muscle.
Two layers: externally – longitudinal; internally – circular.
Innervation:
   Intrinsic – well developed
   Extrinsic – vagus nerve (parasympathetic) and sympathetic
   nerves.

SWALLOWING

Three stages may be described:
• Buccal.
• Pharyngeal.
• Oesophageal.

BUCCAL PHASE

After chewing the mouth is closed. A bolus of food is formed and collected on the upper surface of the tongue. The tongue is moved upwards and backwards by contraction of the mylohyoid and styloglossus muscles. The bolus is forced through the oropharyngeal isthmus. The soft palate is raised, which closes off the nasopharynx. Respiration is reflexly inhibited, this is called deglutition apnoea. The larynx is raised as the bolus passes over it.

PHARYNGEAL PHASE

The tongue now moves upwards towards the pharyngeal wall. The bolus is forced against the epiglottis. The epiglottis is cowl-like and protects the laryngeal opening. The larynx is raised and laryngeal folds close. The bolus of food then passes the lateral edges of the epiglottis in two streams and enters the pharynx posterior to the larynx. The upper oesophageal sphincter or cricopharyngeus muscle (UOS, a thick muscular band 2–3 cm long) relaxes for just about 1 s. The cricopharyngeus contracts which is the start of the peristaltic wave which will drive food down the oesophagus. When the bolus passes the UOS the:
• Larynx is lowered.
• Vocal cords open.
• Epiglottis and tongue resume their resting position.
• Soft palate falls.

OESOPHAGEAL PHASE

At rest, the UOS and lower oesophageal sphincter (LOS) are closed. The LOS is difficult to define anatomically – it is a physiological sphincter (pinch cock) which consists of a high pressure zone 2–6 cm in length. The intraluminal pressure of 10–40 mmHg is normally above intragastric pressure. The barrier is not static.

Neural control of LOS
Relaxation of LOS is controlled by:
- Vagal non-cholinergic, non-adrenergic fibres.
- Activated by a deglutition centre in the medulla.
- LOS relaxes as peristaltic wave passes over it.
  Closure of LOS is controlled by:
- Intrinsic myogenic activity.
- Vagal cholinergic fibres.
- Sympathetic nerves – α-receptor activated.
- Hormones:
  Gastrin
  Motilin – physiological role not established.

Neural control of peristalsis
Peristalsis begins in the pharynx and striated muscle part of the oesophagus, and is regulated by a deglutition centre within the medulla:
- Afferent connections:
  Higher centres – cerebral cortex
  Brain stem – motor nuclei of V, VII, X and XI
  Afferent neurones from the oropharynx.
- Efferent connections – fibres in the vagus nerve.

Intraoesophageal pressure
This is a measure of intrathoracic pressure. It varies with respiration, falls with inspiration and rises with expiration.

Pressure recorded during and after swallowing
Following a single swallow, a single peristaltic wave moves down the oesophagus:
- Recorded as a rise in pressure as the contraction passes over the tip of the catheter.
- Peak pressures 40 mmHg – very variable.
- Velocity of peristaltic wave 2–4 cm/s – faster in upper one-third oesophagus.
- Liquids can enter stomach in advance of the peristaltic wave.
- Solids are pushed down by peristalsis.

Secondary peristalsis
Food which is not cleared by the primary wave distends the oesophagus and initiates another wave – secondary peristalsis.

THE STOMACH

MAJOR FUNCTIONS

The primary functions of the stomach are:
- Temporary storage of food (hopper function).
- Secretes HCl, pepsin and intrinsic factor into the lumen and gastrin into the blood.
- Liquification of food by mechanical and enzyme action.
- Protective action:
  Mucosal protection against exogenous and endogenous aggressors
  Plays a role in vomiting.
- Controls entry of food into the duodenum.

POSITION OF STOMACH IN THE ABDOMEN

The anatomical position of the stomach varies according to:
- Food content.
- Posture.
- Skeletal build.
- Tone of abdominal muscles.
- State of adjacent viscera.
- Typically J-shaped when full and in the standing position.

FUNCTIONAL DIVISIONS OF THE STOMACH

There are two functional divisions:
1. The corpus or body – acts as a food hopper and secretes acid and pepsin.
2. The pyloric antrum – acts as a pump and food mill and secretes an alkaline mucus.

STRUCTURE OF THE STOMACH WALL

The stomach wall is divided into two parts:
1. Mucosa.
2. Muscularis.
   The muscle layer (muscularis) has four components:
1. An external longitudinal layer covered by peritoneum.
2. A circular layer which is thicker in the pyloric antrum and pylorus.
3. An oblique layer lying inside the circular layer, but confined to the proximal stomach only.
4. The muscularis mucosae.

THE MUCOSA

The mucosa is the innermost layer and consists of millions of

tubular glands. The glands reach the mucosal surface through gastric pits and are continuous with a layer of surface mucosal cells. The glands of the mucosa are of two types. In the corpus they are lined by:
- Parietal cells (or oxyntic) secreting HCl.
- Peptic cells secreting pepsinogen.
- Mucus cells secreting mucus.

In the antrum there are no parietal or peptic cells and the glands secrete an alkaline mucus. There are endocrine cells which secrete gastrin into the blood.

INNERVATION

The stomach is innervated by exogenous nerve fibres, from both parasympathetic (the vagus) and sympathetic divisions of the autonomic nervous system. There is also an intrinsic innervation within the stomach wall consisting of two nerve plexuses:
1. The *myenteric (Auerbach) plexus*, lying between circular and longitudinal layers.
2. The *submucous (Meissner's) plexus* lying in the submucosa.

Preganglionic vagus nerve fibres (parasympathetic) relay in both plexuses from which postganglionic fibres arise and innervate smooth muscles and glandular cells. Some sympathetic fibres may also relay here but most do not.

ELECTRICAL ACTIVITY

Two types of electrical activity can be recorded from gastric smooth muscle:
1. Basal electrical rhythm (BER) or electrical control activity (ECA).
2. Electrical response activity (ERA).

Basal electrical rhythm (BER)

BERs are electrical oscillations propagated aborally. They arise from a pacemaker on the proximal greater curvature. They are not associated with contractions.

Electrical response activity (ERA)

ERAs are spike bursts which initiate contractions. They occur in response to the release of acetylcholine by vagal activity. The frequency of ERA is determined by BER.

Vagal stimulation influences the magnitude of contractions but not their frequency.

FILLING OF THE STOMACH ON EATING A MEAL

Food enters the stomach through the gastro-oesophageal opening and enlarges to accommodate the meal by a process known as receptive relaxation, so that between peristaltic contractions the

intragastric pressure remains low. Receptive relaxation is due to reflex relaxation of the corpus and fundus, brought about by vago-vagal reflexes with receptors in:
- Body of the stomach.
- Oesophagus.
- Pharynx.

The transmitters from the efferent vagal fibres are non-cholingeric and non-adrenergic, either purinergic or peptidergic. Reflex relaxation can also be brought about by vago-splanchnic reflexes which are adrenergic.

PHYSICAL PROPERTIES OF THE STOMACH WALL

The fundus and corpus are more compliant than the pyloric antrum, even when exogenously denervated. Tension in the stomach wall increases as it fills, but transmural pressure alters little as the radius increases. The explanation is Laplace's law.

Transmural pressure (P) across the wall of a cylinder is directly proportional to the tension (T) in the wall, and inversely proportional to the radius (R), i.e. $P=T/R$

STOMACH MOVEMENTS AND CONTROL

Once food enters the stomach movements begin and continue until all the contents are emptied into the duodenum. Movements are of two types:
1. Mixing.
2. Propulsive.

Food entering the stomach increases the tension in the walls of the corpus. The increased tension is detected by tension receptors, which initiate and augment contractions in the distal part of the stomach and pyloric antrum. Both afferent and efferent limbs of the reflex reside in the vagus nerves. Contractions form a circular band about 2–3 cm in an axial length and sweep distally towards the pylorus at about 0.5 cm/s, speeding up to 4 cm/s as they approach the terminal antrum. They begin at the pacemaker, and become deeper as the wave moves distally (these are termed peristaltic contractions) (see p. xxx). When the contraction reaches the midantral region the pylorus is open, resistance is low and gastric content flows into the duodenum. As the contraction approaches the pyloric sphincter the latter begins to close, resistance increases so that some gastric content is evacuated but some is retropelled. The sphincter then closes, emptying ceases and powerful antral contractions force the food back into the proximal antrum, grinding the solid food down into finer constituents. Gastric contractions appear about every 20 s in man and usually two or three waves are present at any one time in the antrum. Decreased activity occurs after vagotomy, splanchnic nerve stimulation, adrenergic drugs and atropine. The only physiological stimulus which increases activity is distension of the stomach, other agents increasing activity are insulin hypoglycaemia and cholinergic drugs.

GASTRIC EMPTYING

Gastric emptying depends upon:
- The nature of the meal.
- Liquid meals empty faster than solids.
- Large meals of solid food take longer to empty than smaller ones.
- Meals of coarse foods empty slower than finely ground meals.
- Meals of a high viscosity empty slower than those with a low viscosity.
- Meals of different chemical composition empty at different rates.

CONTROL OF GASTRIC EMPTYING

Gastric emptying is controlled by factors from the stomach, duodenum and outside the abdomen.

Factors arising from the stomach
The volume of the meal – distension of the corpus initiates peristaltic contractions through a vago-vagal reflex. There is a linear relationship between the square root of the volume of the meal remaining in the stomach and the passage of time.

When pressure rises to a critical level in the antrum, activity of the corpus is reflexly inhibited and relaxes the corpus, thus retaining food within the corpus.

Factors acting from the duodenum
Receptors in the duodenum and upper small intestine mostly act as a brake on the stomach.

The neural control of gastric motility by inhibitory reflexes is shown below.

Stimulus	Site of receptor	Site of effector	Pathway	Transmitter
Mechanical	Oesophagus	Corpus	Vago-vagal	NANC
Mechanical	Corpus	Corpus	Vago-vagal	NANC
Mechanical	Antrum	Corpus	Vago-vagal	NANC
Mechanical	Intestine	Antrum	Splanchno-vagal	NANC
Chemical	Upper intestine Excitatory reflexes	Antrum	Uncertain	Uncertain
Mechanical	Corpus	Antrum	Vago-vagal	Cholinergic

These inhibitory reflexes detect the volume and chemical composition of the food and reduce the motility of the pyloric antrum (the gastroduodenal brake) to protect the duodenum from damage either by volume, osmolality or acid. For example, in the case of acid the progression of food along the intestine is at a rate proportional to its neutralisation.

Control is mediated by negative feedback

These negative feedback controls are mediated in part by nervous reflexes and in part by chemical messengers (chalones), but the nervous pathways and the chemical nature of the chalones are not yet fully elucidated. The chemical agents involved have been given the all-embracing name of enterogastrone. Recent work points to cholecystokinin as the important humoral agent released by fatty acids and amino acids as the duodenal regulators of gastric emptying.

Factors from outside the abdomen
Higher nervous centres:
    Fear and anxiety slow emptying
    Excitement increases the emptying rate.

MECHANISMS INVOLVED IN GASTRIC EMPTYING

Both neural and hormonal mechanisms are involved in gastric emptying.

Neural
Vago-vagal reflexes.
Distension of body of stomach reflexly increases activity of distal stomach.
Distension of antrum reflexly inhibits the corpus activity and causes it to relax.
Vago-splanchnic reflexes – in general cause inhibition.
Entero-gastric reflexes through the coeliac and paravertebral ganglia – inhibitory.

Humoral
Excitatory – motilin.
Inhibitory:
    Cholecystokinin-pancreazymin
    Gastric inhibitory polypeptide
    Secretin.

GASTRIC SECRETION

The stomach secretes a fluid, gastric juice. This is the combined secretion of the gastric glands in the fundus, the corpus and the antrum and the composition therefore depends upon the proportions arising from the different cells which make up the glands. The parietal cell secretes hydrochloric acid with the approximate composition:

H⁺ 100 mM

H^+ 100 mM
Cl^- 170 mM
K^+ 10 mM.

Other ions are present in different amounts:
Na^+ 40 mM – with others in smaller concentrations, e.g. Ca^{2+}
Mg^{2+} and bicarbonate ion are also secreted from the antrum.

Other substances in gastric juice are:
Pepsin
Mucus
Intrinsic factor (from parietal cells)
ABO blood group substances
Rennin in the first few weeks of extrauterine life.

MECHANISM OF H⁺ SECRETION

Acid is an important component of the gastric juice. The concentration gradient of H^+ between blood and gastric juice is 10^6. This is established by:

- Active transport of H^+ and Cl^- with both pumps tightly coupled.
- Gastric lumen is –70 mV, negative with respect to blood.
- This PD does not change if diffusion potentials are eliminated.
- The primary reaction is the separation of H^+ and OH^- from water – H^+ is secreted over the luminal membrane and OH^- combines with CO_2 to form bicarbonate.
- The bicarbonate diffuses over the serosal membrane into the blood – this is the alkaline tide (urine alkaline).

The source of the energy comes from oxidative metabolism through ATP. The active hydrogen ion pump can be suppressed by *omeprazole* (a H^+/K^+ ATPase inhibitor), which is of great clinical importance.

CONTROL OF GASTRIC SECRETION

The secretion of gastric juice is the result of both excitatory and inhibitory stimuli.

For the purposes of description it is convenient to classify the control of gastric secretion in phases according to the site at which the stimuli act.

Cephalic phase (stimuli arising from the higher centres of the brain)
Unconditioned reflex – taste (present at birth).
Conditioned reflexes – sight and smell of food (acquired by association with taste).

Gastric phase
Food in the stomach stimulates acid secretion through nervous reflexes and the liberation of the hormone gastrin.

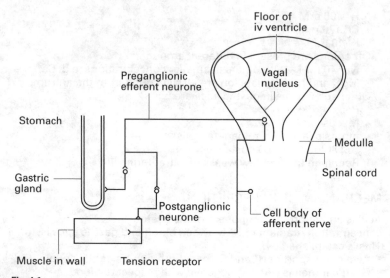

Fig. 4.1
Proposed vago-vagal reflexes controlling gastric motility and secretion from tension
receptors situated in the stomach wall.

Nervous reflexes (Fig. 4.1)

Long vago-vagal reflexes
Distension of the stomach initiates impulses in afferent vagal fibres,
which relay to efferent (preganglionic) fibres in the vagal nucleus,
ending with postganglionic fibres in the enteric plexuses
stimulating acid and pepsin secretion and releasing gastrin from
antral gastric glands.

Short intramural reflexes
I.e. within the stomach wall using the enteric plexuses are
stimulated by distension.
• Those in the corpus are cholinergic, are potentiated by gastrin
  and stimulate acid secretion.
• Those in the antrum release gastrin.

FOOD IN THE STOMACH (Fig. 4.2)

Food in the stomach distends the pyloric antrum and releases
gastrin (through short intramural reflexes) and the products of
digestion, especially of protein, stimulate gastrin release by direct
action on the gastrin cell.

Intestinal phase
The position is less clear here, vago-vagal reflexes and gastrin are

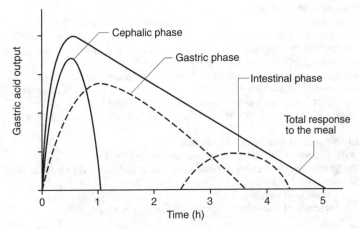

Fig. 4.2
The contributions of the different phases of gastric secretion to the total response of the stomach to a meal.

both thought to be involved, but their effect on gastric secretion is small.

THE ANTRAL HORMONE – GASTRIN

The cell of origin is the G-cell. It has been located by immunohistochemical techniques in the epithelium of the mid-zone of the pyloric glands. The G-cell has basal granules (gastrin) with microvilli, which project into the lumen of the gastric glands and are thought to act as chemoreceptors.

Chemical structure
There are two main naturally occurring active forms, G34 (a chain of 34 amino acids) and G17, and all the activity of gastrin is found in the C-terminal tetrapeptide amide.

Actions of gastrin
The intravenous injection of gastrin produces many effects depending upon dosage. It stimulates:
- Acid and pepsin secretion.
- Intrinsic factor.
- Brunner's glands.
- Pancreatic and biliary secretions.
- Smooth muscle by increasing the tone of the lower oesophageal sphincter and the stomach and small intestine.
- Division and growth of gastroduodenal mucosa and pancreas.
- The release of insulin and calcitonin.

The role of histamine

It has been known for a long time that histamine is a powerful stimulant of acid secretion.

Its action is blocked by H_2 receptor blocking drugs such as cimetidine, but not by H_1 blocking drugs such as mepyramine. H_2 receptor blockers also inhibit secretion in response to food, cholinergic or gastrin stimulation. Histamine is thought to work physiologically by being continuously released by mucosal mast cells, which potentiates the action of acetylcholine and gastrin.

Inhibition of gastrin secretion

Higher centres inhibit gastric secretion and gastric blood flow. These can involve emotions of sadness and fear and sensations of nausea. They act through:
- Reduction in vagal activity.
- Activation of the sympathetic nervous system.

Factors acting in the stomach
Acid secretion sufficient to create a pH of 2.5 or less in the pyloric antrum inhibits gastrin release. Acid secretion can be neutralized by:
- Food, which is quantitatively the most important substance to 'neutralise' secreted acid.
- Non-parietal secretions – bicarbonate can be secreted from the pyloric mucosa.

Factors acting from the small intestine
Just as the duodenal brake can inhibit stomach motility and emptying, it has also been implicated in the inhibition of secretion. This prevents the duodenum from being overloaded with either food or acid. After some gastric operations, food may reach the duodenum at a faster rate and overload it. In this case the patient feels faint and experiences intense nausea. Both neural pathways and humoral agents are involved:
- Neural pathways – afferent fibres in the vagus, efferent pathway unknown.
- Possible humoral agents:
     CCK-PZ (Cholecystokinin-pancreozymin)
     Gastric inhibitory peptide
     GH-RIH (somatostatin).

GASTRIC MUCOSAL BLOOD FLOW

Mucosal blood flow and acid secretion are tightly linked. In the fasting state about 50% of the total gastric blood flow is mucosal. The proportion going to the mucosa increases with increasing rates of secretion. Sympathetic activity and vasoconstrictor drugs reduce mucosal blood flow and secondarily reduce secretion rate.

TESTS OF GASTRIC FUNCTION

These are carried out for the purpose of diagnosing:

- Pernicious anaemia – in this condition there is a total absence of acid secretion.
- Gastric and duodenal ulceration.
- Gastric carcinoma.
- Zollinger–Ellison syndrome – due to a gastrin-producing tumour.

Stomach contents are removed at timed intervals through a nasogastric tube. The volume and acidity of the secretion are measured in the basal state to give the basal acid output (BAO), which normally lies between 1 and 4 mmol/h after stimulation with gastrin (pentagastrin 6 µg/kg body weight). Maximal acid output (PAO_{PG}) lies between 10 and 40 mmol/h). The peak acid output after insulin (PAO_I) is used to test the effectiveness of vagotomy as insulin stimulates efferent vagal fibres through hypothalamic centres, which are activated by a low blood glucose concentration.

PROTECTIVE FUNCTIONS OF THE STOMACH

The gut has defence mechanisms which protect it from external and internal aggressors.

Stomach protection begins in the mouth where the oral sensations can either accept or reject substances the body has learned by experience are either acceptable or harmful. Some irritants which pass the buccal guard can be rejected by vomiting, but the main line of defence is the mucosal barrier.

MUCOSAL BARRIER

A layer of adherent mucus, secreted from the surface epithelial cells about 180 µm thick, covers the whole of the gastric mucosa, with the exception of the gastric pits. It overlies the surface epithelial cells and tightly adheres to them.

The mucous layer forms protection against three of the natural agents which could damage the mucosa:

- Mechanical damage by coarse food.
- Digestion by pepsin.
- Damage by low pH (acid).

Mucus

Mucus is a high-molecular-weight glycoprotein, with a protein backbone and with carbohydrate side chains. Protective properties depend upon its viscogelatinous character conferred by the carbohydrate side chains, as it is water insoluble and adheres to the surface cells. The viscous property provides lubrication protecting it from mechanical damage as the stomach triturates its contents (chyme). Protection against pepsin is provided through a gel-exclusion effect, but pepsin will attack mucin at nonglycosylated regions which are largely devoid of the protective carbohydrate sheath. Mucus degraded by pepsin is replaced by secretion from surface epithelial cells.

Protection against low pH
Defence against acid depends upon:
- Adherent mucus forming an unstirred layer.
- Secretion of bicarbonate into this layer where it meets acid and is neutralised.
- The function of the mucus, which is to prevent the mixing of a relatively small amount of bicarbonate with the bulk acid in the gastric lumen.

Control of mucous secretion
The rate of mucous secretion is increased by:
- Nervous stimulation:
    Vagus nerves
    Sympathetic nerves.
- Acid in the gastric lumen.
- Mechanical and chemical irritation of the mucosa.

VOMITING

Vomiting is coordinated by a bilateral vomiting centre situated near the tractus solitarius and the level of the dorsal vagal nucleus. The factors involved are:

Afferent input
Vagus, chemoreceptor trigger zone, afferents in splanchnic nerves, vestibular nuclei via cerebellum and higher nervous centres.

Efferent output
Somatic: to muscles of chest and abdomen.
Autonomic: vagus and sympathetic outflow.

Stimuli causing vomiting
Central through the chemoreceptor trigger zone (?area postrema).
Peripheral.

Causes of vomiting
These include:
- Cortical: emotion, pain, raised intracranial pressure, migraine.
- Vestibular: motion sickness, labyrinthitis.
- Visceral: mucosal irritation, excessive distension of an abdominal viscus, intestinal obstruction, peritonitis.
- Alterations in body chemistry: diabetic ketosis, renal failure, drug and alcohol ingestion, liver failure.
- Mechanisms:
    Nausea and retching occur as a prelude to vomiting

Contraction of pyloric antrum
Stomach contents pass back into a relaxed corpus and fundus
Deep inspiration, glottis closed, soft palate raised
Violent contractions of the diaphragm and abdominal muscles
ejecting stomach contents
Autonomic disturbances – pallor, sweating, bradycardia,
salivation.

THE EXOCRINE PANCREAS

The pancreas secretes a fluid, the pancreatic juice, that is essential
for the digestion of food. The pancreatic juice is secreted into the
duodenum together with the bile.

STRUCTURE OF THE PANCREAS

The pancreas can be divided into two structural and functional
components:
1. Acinar units – secrete enzyme + some electrolyte.
2. Centro-acinar cells and interlobular ductules – secrete
   electrolytes and water.

EXTERNAL SECRETION

The secretions of the pancreas contains water + electrolytes and
proteolytic enzymes. Together with bile and duodenal juice (mainly
Brunner's gland secretions) these secretions neutralise the gastric
juice to provide an optimal pH of about pH 6.8 for enzyme action.

Water and electrolytes (about 1.5 L/day)
Secretion is iso-osmotic with plasma.
$[Na^+]$ and $[K^+]$ similar to plasma and independent of secretion rate.
$[Ca^{2+}]$ lower than plasma and partly associated with enzyme
secretion.
$[HCO_3^-]$ increases with flow rate, reaching four to five times plasma
levels.
$[Cl^-]$ decreases with increasing flow rate, reciprocally with HCO_3^-.

The pancreatic juice contains enzymes which digest the three main
groups of foodstuffs.

Proteolytic enzymes
These are secreted as inactive precursors of enzymes or zymogens.
Trypsinogen is activated by an enzyme enteropeptidase
(or enterokinase) secreted by the duodenal mucosa to form the
active enzyme trypsin, which plays a key role in digestion as it acts

as a trigger for activating all the other pancreatic zymogens in a pancreatic enzyme cascade.

$$\text{Trypsinogen} \atop \downarrow \quad \leftarrow \text{Enteropeptidase}$$

Trypsin

Zymogen		Enzyme
Trypsinogen	↓	Trypsin
Chymotrypsinogen	↓	Chymotrypsin
Pro-carboxypeptidase A, B	↓	Carboxypeptidase A, B
Pro-elastase	↓	Elastase
Kallikreinogen	↓	Kallikrein
Pro-colipase		Colipase

Amolytic enzymes
Amylase – hydrolyses starch at the α, 1,4 glucosidic bonds→maltose, maltriose, α-dextrins.

Lipolytic enzymes
Lipase – hydrolyses neutral fats to fatty acid and glycerol.

Nucleases
For example – ribonuclease and deoxyribonuclease.

SECRETION OF ENZYMES

Enzymes are synthesised on ribosomes of the rough endoplasmic reticulum (RER) of the acinar cells. They are then passed into the cisternal spaces from where vesicles bud off and carry enzyme precursors to the Golgi complex. There the vesicles coalesce and mature into membrane-bound zymogen granules. During secretion the membranous sacs containing the zymogen granules fuse with apical membrane and the contents are extruded from cell into acinar lumen – exocytosis.

STIMULANTS OF PANCREATIC SECRETION

Pancreatic secretions can be stimulated by hormones and nerve stimulation.

Hormonal stimulation

Secretin
A polypeptide of 27 amino acids.
Produced by cells in the crypts of Lieberkühn.
Released by acid in the small intestine.
Stimulates pancreas to secrete electrolytes and water.

Pancreozymin (identical with cholecystokinin)
Polypeptide of 33 amino acids.

Produced in pear-shaped granular cells (situated among the columnar cells) in the intestinal mucosa.
Apical processes reach into the intestinal lumen.
Stimulates the secretion of pancreatic enzymes and some electrolytes.
Stimulates gall bladder contraction (cholecystokinetic effect).
Plays a role in:
   Gastric emptying
   Satiety.
Note: it is also found at other sites and has a non-digestive function in the body, e.g. in brain where it may be a transmitter.

Gastrin
Because gastrin has the same terminal tetrapeptide as CCK-PZ, it will also stimulate enzyme secretion. Therefore, all the circumstances which release gastrin will stimulate the pancreas.
   These are:
* The direct release of gastrin by food in the pyloric antrum.
* Distension of the antrum by food – mediated by intramural reflexes.
* Distension of the body of the stomach – mediated by a vago-vagal reflex.

Nerve stimulation

The vagus nerve
Acinar cells are innervated by the vagus nerve. These postganglionic vagal neurones release:
* Acetylcholine (ACh) – stimulates enzyme secretion.
* Vasoactive intestinal polypeptide (VIP) – stimulates electrolyte secretion.

CONTROL OF EXOCRINE PANCREATIC SECRETION

For descriptive purposes, control may be divided into three phases according to the site at which the stimuli for secretion act:
1. Cephalic (head end, i.e. psychic).
2. Gastric – stomach.
3. Intestinal – small intestine.

Cephalic phase

Unconditioned reflex
The primary stimulus is taste bringing about a reflex excitation of the vagus:
* Afferent fibres in lingual nerve→tractus solitarius and nucleus→dorsal vagus motor nucleus.
* Efferent fibres in vagus nerves.

Conditioned reflex
The stimuli involved involve sight, smell and sound:
- Afferent neurones in cranial nerves II, VII and VIII to cerebral cortex→dorsal vagus motor nucleus.
- Efferent fibres in vagus nerves. Impulses in vagus pass to:
    The acinar cells
    G-cell of pyloric antrum→release gastrin if pH of antrum is <2.5
    Gastrin stimulates – acinar cells→enzyme secretion – parietal cells of stomach→acid secretion→release of secretion on acidification of duodenum→secretion of electrolytes.

Gastric phase
Distension of stomach with food:
- Elicits a vago-vagal reflex (afferent fibres in vagus→vagal nucleus→vagus nerves).
- Innervates acinar cell→enzyme secretion.
- Innervates gastrin cell→releases gastrin from pyloric antrum:
    stimulates acinar cells→enzyme secretion by the pancreas
    stimulates acid secretion by stomach→acid chyme releases secretin from duodenum→secretion of electrolytes by the pancreas. *Elicits short reflexes* in nerve plexuses of antral wall→gastrin release.

Intestinal phase
The intestinal phase has two components, a neural and hormonal phase.

Neural phase
Vago-vagal reflex initiated by amino acids and fatty acids in the small intestine.
VIPergic (vasoactive intestinal polypeptide)→electrolyte secretion.

Hormonal phase
CCK-PZ→released by amino acids and fatty acids in the chyme stimulates enzyme secretion.
Secretin→released by acid (pH threshold in duodenum 4.5).
Amount released depends on length of gut acidified. Secretin is also released by bile salts.

Neural and hormonal interactions
Secretin potentiates the action of CCK-PZ and acetylcholine. CCK-PZ, ACh and gastrin potentiate the action of secretin.

INHIBITION OF PANCREATIC SECRETION

Inhibitory effects on the pancreas originating from the ileum and colon have been described; these are brought about by fatty acids and hypertonic solutions and are mediated by peptides.

THE SMALL INTESTINE

DIGESTION AND ABSORPTION OF DIETARY CONSTITUENTS

ADAPTATION OF THE SMALL INTESTINE TO ITS DIGESTIVE FUNCTION

To perform its absorptive function efficiently the small intestine provides a great surface area through:
1. Its great length.
2. The provision of spiral disposed folds in the mucous membrane – the *plicae circulares*.

The surface is covered with tiny projections up to about 1 mm in height – the *intestinal villi*. The villi are covered by columnar cells which have on their luminal surfaces *microvilli* (it is the appearance of the microvilli under the microscope which has given rise to the term 'brush border').

The structure of the villus

The villi are finger-like processes covered by columnar cells, of about 20–40 mm^2. Each contains a:
- Central arteriole and venule.
- Communicating capillary plexus.
- Blind-ending lymph vessel – lacteal.
- Smooth muscle strip attached to lacteal – whose contraction empties it.

Each enterocyte may have 1000 microvilli. The tubular glands (crypts of Lieberkühn) open into spaces between villi.

Blood supply
Capillaries are of the fenestrated type – allow for rapid exchanges.
Area exposed 10 m^2.
A counter-current system exists in each villus:
   This prevents very rapid absorption into the circulation
   O_2 crosses from ascending to descending loops by diffusion
   Po_2 of the tip of the villus is low.
Villi contract during digestion through central fibres derived from muscularis mucosae. The villi contract quickly but relax slowly. This action pumps lymph into lacteals of the submucosa.

Enterocyte turnover

Approximately 50–200 g of human gastrointestinal mucosa are renewed daily. The fastest renewal occurs in the ileum (5–7 days). There are three types of cell in the intestinal glands:
1. *Columnar cells – undifferentiated in the crypts:*
   a. migrate upwards along the villus

b. capable of absorptive function
c. associated with enzyme production located in the brush border.
2. *Goblet cells:*
    a. Produce mucus in the crypts
    b. Extrude mucus as they migrate up the villus.
3. *Paneth cells:*
    a. Confined to crypts
    b. Secrete glycoprotein
    c. Function unknown.

In addition there are, scattered throughout the intestine, numbers of endocrine cells, e.g. producing secretin and CCK-PZ.

Intestinal juice

There is fluid in the intestine which does not originate from the pancreas or liver; it is called the *succus entericus*. Two types of gland contribute:
1. *Duodenal (Brunner's) glands found in the submucosa:*
    a. Secrete mucus (protective fluid)
    b. Bicarbonate?
    c. Epidermal growth factor?
    d. Pepsinogen II.
2. *Intestinal glands:*
    a. Present throughout the intestine
    b. Secrete mucus and some enzymes, e.g. enteropeptidase (enterokinase).
    Surface cells also contribute, which secrete bicarbonate.

Control of intestinal secretion

Secretion from Brunner's glands is stimulated by acid chyme and there is evidence that the enteric nervous system and prostaglandins may be involved.

DIGESTION AND ABSORPTION OF CARBOHYDRATES

Although the salivary glands secrete amylase, there is little or no time for amylase to act in the mouth. However, amylase is mixed with the food by the act of chewing and then swallowed, so digestion will continue in the stomach until the acid and pepsin penetrate the bolus of food to bring enzyme action to a halt. It has been estimated that a significant amount of starch can be hydrolysed in the stomach.

Digestion and absorption in the small intestine

Amylase hydrolyses starch at α, 1,4 glucosidic bonds, resulting in compounds such as maltose, maltriose and α-dextrins. The final breakdown to monosaccharides is brought about by brush border enzymes on the surface of the epithelial cells:

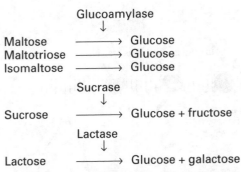

Monosaccharides (glucose, galactose and fructose) are largely absorbed in the duodenum and upper jejunum. The rate of glucose absorption is carrier mediated.

Na$^+$-coupled glucose transport
The carrier, a symport, has glucose attached to one port and Na$^+$ to the other. As the Na$^+$ concentration in the cell is low and the membrane potential negative ($\cong$ –70 mV), it enters down its electrochemical gradient allowing glucose and galactose to utilise this energy and enter the cell. Glucose and galactose then diffuse passively out of the cell down their concentration gradients over the basolateral membrane into the capillary of the villus. The movement of glucose and galactose is then said to occur by secondary active transport. Na$^+$ leaves the cell through the action of the sodium pump at the basolateral membrane, keeping the cell [Na$^+$] low.

THE DIGESTION AND ABSORPTION OF PROTEINS

Protein digestion begins in the stomach where pepsin acts as an endopeptidase and hydrolyses the large protein molecules at bonds away from the carboxy and amino terminus to near aromatic groups to form shorter peptide chains. The major site for protein digestion is the small intestine under the influence of the pancreatic peptidases trypsin and chymotrypsin, which hydrolyse proteins into smaller molecules up to about six residues long. In the small intestine *carboxypeptidase* removes amino acids in sequence from the carboxyl ends of molecules. Short chain peptides are hydrolysed to amino acids at the brush border, the enzymes involved are:

aminopeptidase→splits off amino acids from the amino terminal
dipeptidase→hydrolyses dipeptides into two amino acids.

Amino acid transport
The end products of protein digestion are amino acids. More than 90% of exogenous and endogenous protein is absorbed in this form. Absorption is rapid and stereospecific – L-isomers are actively

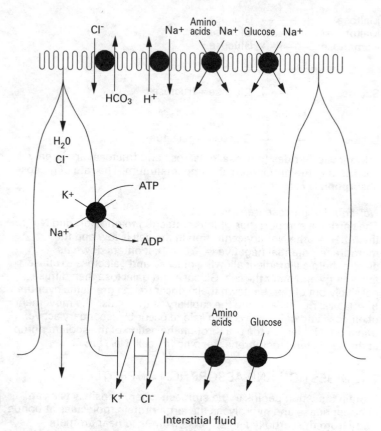

Fig. 4.3
A theroetical model of isosmotic intestinal transport showing the coupled transport of glucose and amino acids to sodium, hydrogen : sodium exchange and chloride : bicarbonate exchange at the luminal cell membrane.

transported. The mechanisms involved are carrier mediated and sodium dependent (as for glucose). Energy is obtained from secondary active transport (similar to glucose). Some small peptides are transported into the cells by a different mechanism along with H^+ ions. Amino acids leave the cell at the basolateral border by diffusion or carrier-mediated transport.

Coupled transport of amino acids and glucose is shown in Figure 4.3.

Carrier mechanisms for amino acids
The mechanisms which transport amino acids can be divided into four classes:
1. Neutral.
2. Basic.
3. Acidic.
4. Imino acids.

Absorption of larger protein molecules
In neonates, antibodies and other proteins in the mother's colostrum may be absorbed intact by pinocytosis.

DIGESTION AND ABSORPTION OF FAT

The intake of fat varies considerably but on average is from 60 to 100 g/day. Digestion is an efficient process with only about 5 g being lost per day in the faeces.

Dietary fat
Triglycerides (90%)
Phospholipid
Cholesterol and its esters.

Emulsification
As fats are water insoluble they have to be put into a form suitable for digestion. Fat passes into the duodenum as oily drops in the chyme. CCK-PZ is released into the blood. This activates contractions in the gall bladder, pushing concentrated bile into the duodenum. The pancreas is stimulated to secrete enzymes, including lipase and colipase. The detergent action of bile salts reduces the surface tension, thus emulsifying the fat. The bile salts aggregate in groups called micelles.

Hydrolysis of fat
Lipolysis leads to the production of monoglycerides and fatty acids. Bile salts accumulate at the fat–water interface and would inhibit further action but for a co-factor secreted in the pancreatic juice called *colipase* (a protein). Colipase attaches to fat droplets and to bile salt micelles to form a binding site for the lipase. Further hydrolysis occurs, with the incorporation of the products into the bile salt micelles to form mixed micelles.

Absorption of fat
Mixed micelles reach the enterocyte surface membrane, and the monoglyceride and fatty acids, being fat soluble, pass into and across the cell membrane. Within the cell they are re-esterified either with glycerol, through a glycerophosphate pathway, or direct acylation of 2-monoglyeride. Within the endoplasmic reticulum chylomicrons are formed (through a combination of triglyceride,

phospholipid and apoprotein). They pass to the Golgi complex and sugars are added to the apoprotein by glucosyltransferases. The chylomicrons leave the cell by exocytosis into the interstitial space. From there, they are taken into the lymphatics back to the thoracic duct and into the blood stream. Chylomicrons are 0.5–1 μm in diameter and after a fatty meal cause the plasma to become cloudy.

ABSORPTION OF OTHER SUBSTANCES

Vitamins

Fat-soluble vitamins A, D, E and K and their precursors
These are incorporated into micelles and therefore depend upon the presence of bile salts for their absorption.

Water-soluble vitamins
Some, e.g. thiamine (B_1 and ascorbic acid (C)) have a sodium dependence similar to glucose and amino acids. Others may undergo chemical change, e.g. folic acid undergoes reduction and methylation to methyl tetrahydrofolate.

Bile salts
These are absorbed in terminal ileum by specific transport mechanisms. Vitamin B_{12} binds to glycoproteins (intrinsic factor) secreted mainly by parietal (acid producing) cells:
• Resists enzymatic digestion.
• Absorbed in terminal ileum after attachment to specific receptors.
• Mucosal uptake by pinocytosis – facilitated by Ca^{2+} or Mg^{2+}.
• Intrinsic factor degraded in epithelial cell.
• Enters portal blood bound to a transport protein transcobalamin II.

ABSORPTION OF WATER AND ELECTROLYTE

Normal turnover of salt and water

Average volumes of fluid (litres) entering the duodenum per day

In the diet	2.0
Saliva	1.0
Gastric juice	1.5
Bile	1.0
Pancreatic juice	1.5
Total	7 L/day

Sites of absorption

Small bowel ≅ 6.0 L/day
Large bowel ≅ 1 L/day
In faeces ≅ 0.1–0.2 L/day

Transport of water
Transfer into blood is passive and is secondary to solute transport moving down an osmotic gradient created by solute transport.

Electrolytes

Sodium
About 90% of the sodium is absorbed in the small intestine, with about 10% in the large intestine. There are five ways in which Na^+ is absorbed:

1. Passive movement into the enterocyte with active transport out of the cell across the basolateral membrane.
2. Absorption of Na^+ occurs in association with glucose and amino acids (see above).
3. Some Na^+ is entrained to water movement through the intercellular spaces – solvent drag.
4. Some passes across the apical membrane through a co-transport system with chloride and actively out over the basolateral membrane (through the sodium pump). Chloride moves out passively down its concentration gradient.
5. Some Na^+ is absorbed in exchange for H^+. The movement of electrolyte into the intercellular space creates the osmotic gradient for water to pass through cell junctions which are very leaky.

Potassium
Potassium transport is mostly passive, down concentration and electrical gradients. This occurs by solvent drag in the intercellular space.

Bicarbonate
Bicarbonate ions are absorbed:
- In exchange for chloride ions.
- As part of the Na^+/H^+ exchange and as CO_2. This occurs as a result of the conversion of:

$$H^+ + HCO_3^- = CO_2 + H_2O$$

Chloride
This occurs:
- By passive diffusion.
- As part of the $Na^+ - Cl^-$ co-transporter (see above).
- By solvent drag.

Calcium absorption
See Chapter 10.

Iron absorption
See Chapter one.

SMALL INTESTINAL MOTILITY

The efficient digestion and absorption of food demands that it is

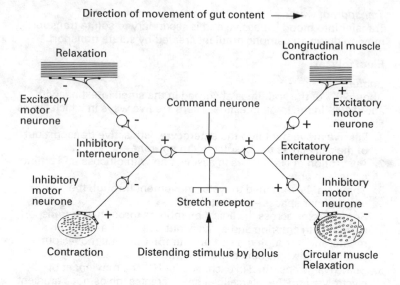

Fig. 4.4
The myenteric reflex. A schematic diagram illustrating the simplest neuronal circuitry in the myenteric plexus which would allow for contraction of the gut behind a distending stimulus from a food bolus and relaxation in front to account for onward progression of chyme. Units such as this would have to be replicated down the whole length of the gut.

moved through the gastrointestinal tract in an orderly controlled fashion. This is carried out by the contractions of the smooth muscle coats of the gut, to bring about:

- Net aboral movement of gut content.
- Circulation of the content to bring the products of digestion into contact with the absorptive surfaces.

Also see Figure 4.4, which illustrates the myenteric reflex.

The motility of the small intestine depends upon the integrated action of myogenic, neural and hormonal mechanisms. Two types of movement are recognised:

1. Propulsive.
2. Non-propulsive or mixing movements.

Propulsive movements
Peristalsis – ring-like contractions moving distally (at about 10–15 cm/min).
Controlled by local reflexes in the enteric plexuses.
Stimulus is distension by a bolus of food which causes:

Contraction of the gut above the bolus
Inhibition below.

Non-propulsive (mixing movements)

Pendular movements – shunt chyme up and down segments of the
intestine controlled by cholinergic reflex arcs.
Segmentation – alternate contraction and relaxation of gut
segments thought to be myogenic.

INTESTINAL MOVEMENTS DURING FASTING

Interdigestive motility occurs in cycles. Each cycle passes from
duodenum to terminal ileum in about 100 min and then recycles.
The contractions are intense (about 10–20/min) and forceful. The
activity is called the migrating myoelectrical complex (MMC)
(Fig. 4.5).

Three phases are described:
Phase I – quiescence.
Phase II – intermittent activity.
Phase III – irregular activity (lasts about 10 min).

The function of phase III is to sweep the remnants of the previous
meal into the colon. The MMC is disrupted within minutes of taking
a meal.

ELECTRICAL ACTIVITY

The intrinsic basal electrical activity arises from a pacemaker in the
duodenal wall. Electrical waves are propagated down the intestine
at different rates in different regions:
- Rate in duodenum 12–16/min.
- Rate in ileum 6–8/min.

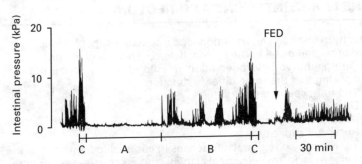

Fig. 4.5
A migrating myoelectric complex (MMC), recorded from the jejunum of the dog
manometrically, and initiation of the fed pattern of motility. A = Phase I; P = phase II; C
= phase III (activity front). The animal was fed at the arrow, initiating the fed pattern.
(Clark and colleagues, unpublished observations.)

Myogenic electrical activity is the basis for all intestinal movements. It can be modified by:
- Cholinergic nerves (vagus)→excitatory effects
- Adrenergic nerves (sympathetic)→inhibitory effects
- Non-cholinergic non-adrenergic (purinergic and peptidergic)→either excitation or inhibition.

Neural modulators of myogenic electrical activity
5-Hydroxytryptophane, encephalins, endorphins, dopamine, vasoactive intestinal polypeptide.

Hormones
Gastrin, secretin, motilin, CCK-PZ.
The transit time from the stomach to the ileocaecal valve $\cong$ 3 h.

THE ILEO-COLIC JUNCTION

This is a region approximately 4 cm long which is responsible for raising intraluminal pressure ($\cong$ 20 mmHg). It functions as a sphincter. It is closed for most of the time to prevent reflux from colon into ileum. As a peristaltic wave approaches it relaxes, a process which involves local mechanisms within the gut wall. Thus it is unaffected by extrinsic denervation. Under in vitro conditions this region:
- Contracts to acetylcholine and α-adrenergic stimulation.
- Relaxes to β-adrenergic stimulation.
The ileo-colic junction is also affected by hormones:
- Contraction by secretin.
- Relaxation by gastrin.

THE LARGE INTESTINE AND RECTUM

The human colon can be subdivided into two distinct regions, proximal and distal to the splenic flexure. These two regions have differing embryology, innervation and blood supply, so it is not surprising that they also differ in function:
The right colon has a greater:
- Diameter
- Absorptive capacity
- Reservoir capacity.
There are also regional differences in contractile frequency of the circular muscle, myoelectrical activity, and responsiveness to both physiological and pharmacological stimuli.

Function
Stores undigested remnants prior to defaecation.
Transports these remnants into the rectum for discharge.
Plays a role in salt and water balance.

Contains bacteria that break down complex carbohydrates and synthesise vitamin K.

MOVEMENTS OF THE LARGE INTESTINE AND RECTUM

The caecum and colon fill from the small intestine when a meal is taken due to a gastro-ileal reflex. As peristaltic waves approach the ileo-colic sphincter it opens, allowing onward progression of gut content. Food enters the caecum about 4 h after a meal, reaches the hepatic flexure in 6 h and splenic flexure in 9 h. After approximately 12 h it has reached the pelvic colon. The transit rate is variable and is faster when the diet contains indigestible fibre.

COLONIC MOTILITY

The colon can produce different types of contractions, non-propulsive and propulsive.

Non-propulsive
Segmental contractions (of circular muscle) of low frequency (3–4/min) cause deep infoldings of the wall, forming pouches called *haustra*. The circular muscle relaxes only to contract again in a new area mixing the content and bringing it into contact with the mucosa, thus allowing absorption to take place. Pressures of between 10 and 60 mmHg have been recorded during contractions. Pressure increases after the ingestion of a meal and is diminished during sleep.

Propulsive
Peristalsis
Mass movements occur three to four times a day as strong peristaltic waves which push the gut content from proximal, through the transverse, and descending colon to the sigmoid colon. Haustral folds disappear before the peristaltic wave begins and reforms afterwards. By this time the content is usually more solid as water and electrolyte have been absorbed.

Control of colonic movements
Regional differences in phasic motor activity occur in response to food. The proximal colon responds rapidly with a maximal response which decreases, whereas the distal colon responds more slowly and activity is more sustained. The colon responds to feeding with two phases:
1. Cephalic
2. Intestinal.
   Both phases are modulated by nervous and hormonal mechanisms. On the nervous side there is evidence of a cephalic phase and a response to the ingestion of food. The latter has been termed the gastro-colic reflex, but this a misnomer as it also occurs

in the absence of a stomach. A vago-vagal reflex from the colon itself has also been demonstrated in animals. Dietary factors, particularly fat, have been implicated, suggesting that CCK-PZ may also be involved.

DEFAECATION

The act of defaecation is mediated by voluntary action associated with reflexes through the sacral cord. The contents of the sigmoid colon are moved into the rectum by peristalsis. Tension in the rectal wall increases and this is signalled to the brain as a desire to defaecate. If it is socially undesirable to defaecate, the sensation can be suppressed and contraction of the anal sphincter occurs. If conditions are appropriate:

- Tension in the rectum leads to further reflex contraction of the sigmoid colon and rectum.
- The nervous pathway is via the pelvic splanchnics.
- Reflex inhibition of the internal and external anal sphincters then occurs.
- Reflexes are reinforced as contents pass through sphincters.

To aid in the passage of the faeces a subject can voluntarily increase the intra-abdominal pressure. This is induced by:

- Expiration against a closed glottis.
- Contraction of the abdominal muscles.
- Descent of the diaphragm.

In the final stages the levator ani contracts to straighten out the anorectal angle. This pulls up the anal canal over the faecal mass. The reflexes are reinforced as the faeces causes tactile stimulation of anal canal and skin. The length of colon emptied varies from the whole left side of the colon to the rectum not being completely evacuated.

The frequency of defaecation is very variable and determined by diet and habit. It can occur as frequently as three times a day or as little as three times a week.

ABSORPTION AND SECRETION IN THE LARGE INTESTINE

The colon absorbs electrolyte and water to such an extent that in health the normal stool is solid.

Substance	Amounts entering colon	Amounts in stool
Water (ml)	500–1500	100
Na^+ (mmol)	210	5–9
K^+ (mmol)	9	5–17
Cl^- (mmol)	105	2
Bicarbonate (mmol)	75	? because of loss as CO_2

Sodium absorption is active and generates a PD of 30–40 mV across the mucosa. Chloride is absorbed in exchange for bicarbonate. Potassium is, in part, actively secreted and, in part, exchanged for sodium. Mucus is secreted. It acts as a lubricant and has a pH of approximately 8.0.

BACTERIA IN THE INTESTINE

Bacteria are found naturally in the intestine. In the newborn there are no bacteria but colonisation occurs after a few days. The bacteria in the intestine form a useful symbiosis. For example, in the premature baby there is no vitamin K produced in the gut.

In the adult different bacteria inhabit different sections of the gastrointestinal tract.

Stomach
Acid in the stomach protects against a massive invasion of bacteria. However, in a proportion of the population a bacterium called *Helicobacter pylori* infects the stomach mucosa. This is responsible for some cases of indigestion and peptic ulceration.

Small intestine
Acid in the stomach protects against a massive invasion. Small intestinal fluid contains 10^3–10^5 organisms/ml. These are:
Streptococci
Staphylococci
Lactobacilli
Fungi.

Colon
Predominantly:
Bacteriodes
Bifidobacteria
Streptococci
Lactobacilli.

EFFECTS OF INTESTINAL BACTERIA

The bacteria in the intestine subserve different functions:
Synthesis of vitamin K
In germ-free animals (gnotobiotics) a lack of bacteria can lead to the following:
• Rodents develop a large caecum.
• Villi are slender and the total absorptive area reduced.
• Turnover of cells halved.
• Lymphocytes and plasma cells in the wall are reduced.
• Mucosal connective tissue is thinner.

- Mucus accumulates.
- More proteins are lost in the faeces.
  Urea ($\cong$ 7 g/day) is degraded in colon to NH_3, which is re-absorbed via the liver. Products of bacterial metabolism are absorbed and secreted in the urine (e.g. phenol, cresol, indol, skatol, pyrocatechol).

GAS IN THE INTESTINE

Gas in the intestine can come from a variety of sources:
- *Gastric gas* – swallowed air.
- *Colonic gas* – variable in amount (400–1200 ml) and composition:
  nitrogen 25–80%
  oxygen 0–2.5%
  hydrogen 0.5–27%
  carbon monoxide 0–26%
  carbon dioxide 5–29%.
- *Other gases present may be*:
  methane
  hydrogen sulphide.
  The amounts of gas produced depend upon diet. For example, certain foods, e.g. beans, cabbages, onions, cauliflower and corn, which contain large amounts of non-absorbable carbohydrate, can be fermented by colonic bacteria and are the cause of increased amounts of flatus.

THE FAECES

In the newborn the stool passed on first day of life (meconium) has unique properties:
- Viscid
- Odourless
- Mainly carbohydrate.
  In the adult the faeces have the following usual characteristics:
- Stool weight 100 g/day (Europeans).
- pH = 7.0–7.5.
- Colour due to stercobilin.
- Faecal odour due to indole and skatole.
- Water content 60–80%.
- Faeces are also formed during starvation.
  The faeces are normally composed of residues of:
- Mucus, bile, leucoctyes, desquamated cells, enormous numbers of bacteria.
- Inorganic material 10–20% (calcium and phosphates).
- Nitrogenous material 1–2 g N/day (non-dietary).
- Fat 5 g/day (non-dietary).
- Fatty acids 10–20% of total solid.

Importance of fibre and roughage
When sufficient roughage is present in the diet the bulk of the

faeces is increased and the transit time reduced. A high fibre diet reduces the incidence of:

- Colonic cancer
- Coronary heart disease
- Other diseases.

HEPATIC AND BILIARY FUNCTION

THE LIVER

The liver is a major organ with multiple functions. It is involved in the synthesis of plasma proteins, the formation and secretion of bile, metabolism, storage of key substances and the removal of antigens absorbed by the gastrointestinal tract.

The liver is particularly involved in the metabolism of:

Amino acids
Carbohydrate
Ethanol
Hormones
Drugs.

The liver is an important store of:

Glycogen
Vitamins
Fe and Cu.

Circulation through the liver

About 30% (1500 ml/min) of the cardiac output enters the liver from two sources:

Hepatic artery
Mean pressure = 100 mmHg.
500 ml/min, at a mean oxygen tension of 95 mmHg and 97% saturated.

Portal vein from the splanchnic vascular bed
Mean pressure = 8 mmHg.
1L/min, at a mean tension of 60–75 mmHg about 70–60% saturated.

Blood from both sources is thoroughly mixed within the liver sinusoids and is collected in the central vein and enters the vena cava through the hepatic vein. Hepatic blood flow increases after a meal and decreases during exercise.

Synthetic functions of the liver

One of the major functions is the synthesis of plasma proteins. These include:

1. Albumen – 150–200 mg/kg/day is synthesised with 6–10% of this pool degraded each day.
2. Globulins – clotting factors:
    Fibrinogen
    Pro-thrombin
    Factors V, VII, IX and X
    Fibrinolytic factors.
3. Transport proteins:
    Thyroid binding globulin
    Sex hormone binding protein
    Transcortin
    Transferrin
    Caeruloplasmin (Cu)
    Protein moiety of plasma lipoprotein.
4. Enzymes for intermediary metabolism.
5. Plasma enzymes:
    Lecithin – cholesterol acyl transferase
6. Enzyme inhibitors:
    α_1-anti-trypsin.

Detoxification

The liver has important functions in the removal and inactivation of substances that are toxic. Many drugs are metabolised in the liver. The stages in drug metabolism include:

Phase 1
Oxidation
Reduction
Hydrolysis.
Phase 2
Conjugation with:
  Glucuronic acid
  Glycine
  Sulphates.

After conjugation the products are excreted in the bile and lost in the faeces. Some are returned to the circulation to be eventually lost after filtering by the kidney in the urine.

Bilirubin

When erythrocytes reach the end of their lifespan they are taken up by reticuloendothelial cells. There, the haemoglobin is broken down into Fe, globin and haem. The haem moiety cannot be recycled and has to be excreted. Haem is converted to bilirubin which is then bound in the plasma to albumen. This complex is then carried to the liver where it is separated from the albumen and conjugated. It is then secreted into the bile as the glucuronide.

Enzyme induction

During long-term administration of drugs metabolic enzymes are induced to increase drug removal. For example, ethanol and barbiturates have this effect. The enzymes induced are non-specific.

Inactivation of hormones

The liver has an important role in the inactivation of hormones such as:
Insulin
Glucagon
Cortisol
Aldosterone
Testosterone
Thyroid hormones
De-iodination of thyroxin T_4 to T_3.

Metabolism of ethanol

Approximately 2–10% of ingested ethanol is excreted unchanged in the breath and urine. The liver is involved in the metabolism of ethanol and uses three metabolic pathways:
1. Involving alcohol dehydrogenase:

$$CH_3 \cdot CH_2 \, OH + NAD^+ \rightarrow CH_3CHO + NADH + H^+.$$

2. Utilising the mixed function oxidase system of the smooth endoplasmic reticulum:

$$CH_3 \cdot CH_2 \, OH + NADPH + H^+ + O_2 \rightarrow CH_3CHO + NADP + 2H_2O.$$

3. Using hydrogen peroxidase (accounts for only 1%):

$$NADPH + H^+ + O_2 \rightarrow NADP + H_2O_2$$
$$H_2O_2 + CH_3 \cdot CH_2 \, OH \rightarrow CH_3CHO + 2H_2O.$$

Acetaldehyde (CH_3CHO) is further oxidised by mitochondria to acetic acid using aldehyde dehydrogenases.

$$CH_3CHO + NAD^+ + H_2O \rightarrow CH_2COOH + NADH + H^+$$

Storage function

The liver acts as a store for:
Glycogen
Triglyceride
Vitamins:
  A, D, E and K
  Riboflavin
  Nicotinamide
  Pyroxidine
  Folic acid
  B_{12} (up to 4 years' supply)
  Metal ions (Fe in Kupfer cells, Cu).

Antigen clearance

Antigens, including bacteria, are removed from the portal blood.

BILIARY SYSTEM
Composition of bile
Bile is iso-osmotic with plasma (300 mosmol/kg) and has a pH of between 5.7 and 8.6. The major components are:

Organic components	Concentration (mg/100 g)	% of total solids
Bile acids	140–2230	8–53
Phospholipid	140–810	9–21
Cholesterol	100–320	3–11
Bile pigments	10–70	0.4–2
Protein	30–300	1.3–10

Inorganic components	Concentration (mmol/l)
Na	146–165
K	2.7–4.9
Ca	5.0–9.6
Mg	2.8–6.0
Cl	88–115
HCO_3	27–55

Bile acids
Bile acids are steroids derived from cholesterol. The primary bile acids are secreted conjugated with either taurine or glycine as Na salts. They include:
Cholic acid
Chenodeoxycholic acid.

Properties of bile salts
Bile salts are detergents with a hydrophobic steroid nucleus and a hydrophylic side chain with hydroxyl groups. Groups of molecules (10–12) are oriented with the steroid nucleus on the inside and the polar chains on the outside to form a micelle.

Secondary bile acids
The secondary bile acids include:
- Cholic acid which in the presence of intestinal bacteria become deoxycholic acid.
- Chenodeoxycholic acid which can be converted to lithocholic acid.
- Cholesterol which is insoluble in water and held in solution as a mixed micelle with bile salts.

Bile secretion
Bile flow occurs as a result of active secretion of solute as a result of a local osmotic gradient. Water follows the solute passively. The total biliary flow is the sum of two components.

1. Bile acid-dependent flow where the secretion rate is proportional to the secretion of bile acids (choleresis).
2. Bile acid-independent flow. This is defined as the flow calculated in the absence of any bile acid-dependent flow. This process is Na dependent.

The composition of bile is modified during its passage down the biliary tree. HCO_3 and H_2O are secreted in response to secretin. Na, Cl, HCO_3 and H_2O are absorbed in the gall bladder such that there is a 10-fold concentration. However, the bile remains isotonic as a consequence of micelle formation. At the same time bile salts, bile pigments, cholesterol and phospholipids are concentrated. The total output of bile is approximately 500–600 ml/day.

Entry of bile into the duodenum
This is controlled by the reciprocal activity of the gall bladder and choledocho-duodenal sphincter.

Gall bladder contraction
Between meals the choledocho-duodenal sphincter contracts and the biliary pressure diverts bile into the gall bladder. As the gall bladder fills there is an active relaxation and at the same time there is a reduction in bile volume as a consequence of active concentration. The gall bladder contracts when food is in the upper intestine. This is activated by a release by CCK-PZ from the mucosa and potentiated by secretin and vagal activity.

Control of the choledocho-duodenal sphincter
The sphincter resistance is increased by antral contractions and acid in the duodenum. These changes may involve cholinergic mechanisms.

The sphincter resistance is decreased by CCK-PZ, vagotomy and atropine. Vagotomy and atropine interfere with cholinergic mechanisms.

Function of bile acids
Bile acids have the following functions:
- Inhibition of cholesterol 7α-hydroxylase in the liver which controls the rate of synthesis by negative feedback.
- Stimulate bile acid-dependent mechanisms – choleresis.
- Increase secretion of phospholipids.
- Solubilise cholesterol by incorporation into micelles.
- Aid in the emulsification of fats with phospholipids, fatty acids and monoglycerides – detergent action.
- Aid absorption of fat-soluble vitamins.
- Stimulate pancreatic secretion by releasing CCK-PZ.
- Inhibit pancreatic lipase.
- Inhibit electrolyte and water absorption in the colon.
- Increase colonic motility.
  Very little bile acid reaches the colon.

The enterohepatic circulation

90% of the bile acids are actively absorbed in the terminal ileum. Cholic acid is absorbed more effectively than chenodeoxycholic acid and the conjugated acids are absorbed better than the unconjugated acids. There is a re-excretion from the liver after returning via the portal blood. Approximately 500–700 mg are lost per day in the faeces. This is replaced by hepatic synthesis. There is a total bile acid pool of 2–4 g in the liver. 20–30 g/day of bile acid enters the duodenum, which means that the bile acid pool recycles 6–10 times in 24 h.

5. Food, energy and nutrition

THE COMPOSITION OF FOODS

In the following sections the nature and components of everyday foods are described.

Milk
Milk is a complete food. It is the sole natural food which provides nutrition for infants up to 3 months old.

Composition of human milk
Protein (1.3%):
  Casein
  Whey protein – α- and β-lactalbumin.
Fat (4.1%).
Carbohydrate – lactose (7.2%).
Minerals:
  Calcium (25–35 mg%)
  Phosphate (10–20 mg%)
  Iron – a poor source (0.1–0.2 mg%).
Vitamins:
  A and B group satisfactory source
  C 2–5 mg/100 ml
  D poor source.
Water 87%.
Energy value 298 kJ/100 g.

Cows' vs. Human milk

	Carbohydrate	Fat	Protein	Minerals	Energy
Human	6.8 g	4.0 g	1.5 g	8.2 (meq)	285 kJ
Cows'	5.0 g	3.5 g	3.5 g	29.8 (meq)	275 kJ

Cows' milk contains more protein, calcium, phosphate, sodium and potassium, whereas human milk has more carbohydrate.

Milk and milk products
Raw milk contains microorganisms. It provides a suitable medium for bacterial growth.
  Pathogens can be removed by heat by processes of:
• *Pasteurisation* – milk is subjected to a high temperature for a short time (HTST) 71–73°C for 15 s.

- *UHT milk* – ultra heat-treated (UHT) 130–150°C for 1–2 s.
- *Sterilised milk* – homogenised milk is bottled then heated to about 120°C for 20–60 min.

Modifications to milk
Dried milk – spray or roller dried.
Condensed milk:
    Water removed by vacuum
    Sterilised by high temperature in tins.
Cream – fat separated by centrifugation (40–50% fat).
    Half cream 12% by weight milk fat
    Single cream 18%
    Whipping cream 45%
    Double cream 48%
    Clotted cream 55%.
Skimmed milk – milk after removal of fat.
Semi-skimmed milk – 1.5–1.8% fat by law.
Butter – made by prolonged shaking of cream to make fat droplets coalesce (60% saturated fat).
Cheese
    Milk proteins are coagulated with rennet at 30°C
    Fat is trapped in the coagulated mass, which is pressed and ripened by bacterial and mould action.
    Good source of protein, fat, and calcium.

Margarine

Margarine is made by the hydration of vegetable oil and animal fat. Soft margarine contains:

- 25–50% polyunsaturated fatty acids.
- Vitamins A, D and E are sometimes added.

Meat

Meat is skeletal muscle, liver and kidney. It is rich in nuclear protein. Liver and kidney contain less fat than skeletal muscle. Liver is rich in vitamins A and D and iron.

Fish

Fish is an important source of protein. White fish contains about 2% fat. Herring and mackerel (fatty or oily fish) contain 5–18% fat. It is rich in vitamins A and D and two helpings of fatty fish a week help to protect from coronary heart disease (CHD).

Eggs

This is a highly nutritious food, and a good source of protein, vitamins and minerals. An egg contains (per 100 g):

- Protein (11.8 g) – used as a reference protein biological value 100 (see p. 164).
- Fat (9.6 g).

- Minerals:
    Ca 50 mg; Fe 2.0 mg.
- Vitamins:
    Retinol 140 µg
    Vitamin D 1.75 µg
    Thiamine 0.09 mg
    Riboflavin 0.47 mg.

Cereals
The main constituents are:
- Starch 70%
- Protein 11% – gluten (gliadin and glutelin)
- Fat 0.5–8%
- Ca, P, Fe 2%
- Water 15%.
    The cereal grain consists of:
- Endosperm (centre of wheatgrain):
    Outer layer – proteins glutelin and gliadin; nicotinic acid
    Inner layer – starch and protein
- Germ is rich in vitamins of the B group.
    Cereals can be prepared in different ways:
- *Whole meal* (92–100% extraction) – contains a large proportion of indigestible fibrous matter, vitamins and minerals.
- *White flour* (72% extraction):
    Better keeping properties than whole meal
    Loss of some vitamins, proteins and minerals.
    In Britain it is fortified by adding calcium, iron and thiamine.
- *Wheat meal flour* (85% extraction) – used in manufacture of some brown bread, retains some of the B vitamins, fibre and minerals.
Note: that the intestinal mucosa of patients with coeliac disease is damaged by the action of the α-gliadin of wheat, causing villous atrophy and malabsorption.

Maize
Main protein is zein and is devoid of tryptophan. Niacin (nicotinic acid) is present in bound form and only released when food is prepared with alkali, e.g. in the preparation of tortilla.

Green vegetables –
Most green vegetables are good sources of:
- Dietary fibre.
- Ascorbic acid, carotene (precursor of vitamin A) and minerals.

Fruits
- Usually of little calorific value.
- Valuable sources of vitamin C (ascorbic acid).
- Contain variable amounts of fructose and sucrose when ripe.
- The banana is unique in supplying starch as well.

Potatoes
Consist of:
- Mainly starch.
- Ascorbic acid (about one-quarter of intake from this source).
- Small amounts of vegetable protein.
- Fe and F.

Pulses (beans, peas and lentils)
Consist of:
- Proteins – 20–25% of the dry state; complement cereals nutritionally in essential amino acids.
- Ascorbic acid and carotene, but only in the fresh state.
- Indigestible polysaccharide.
  Soya bean and ground nuts have a high fat content and are rich in linoleic acid.

Chocolate and cocoa
These have high energy values.
- Protein.
- Carbohydrate.
- Fat (may be as high as 30% or more).
- Fe in cocoa (10.5 mg/100 g powder).

Alcohol
Alcohol can provide energy.
- Energy value 1 g provides 7 kcal or 29 kJ/g (1 ml of pure alcohol weighs 0.79 g).
- Rapidly absorbed from the gut
- Metabolised by the liver
- Some lost in urine and the breath.
  Alcohol has many other effects on the body:
- Blood alcohol concentration falls 3 mM/h.
- 50 mM (7 g) metabolised/h.
- Inhibits gluconeogenesis.
- Lowers blood sugar.
- Potentiates insulin release by glucose.
- Releases inhibitions and progressively induces motor incoordination.
- Deep coma occurs at blood levels of 90 mM.
- At certain levels it prevents aggregation of platelets and may protect against CHD.

Beer
Made by fermentation by yeast of a solution of carbohydrate (usually derived from barley). Beer contains:
- Usually 4–6% alcohol.
- Some carbohydrate (beer stains are sticky on a table).
- B group of vitamins from the yeast.
  In some individuals contributes significantly to energy intake.

ENERGY REQUIREMENTS AND ENERGY EXPENDITURE

UNITS OF ENERGY

1 kilocalorie (kcal) = 4184 joules (J) or 4.184 kilojoules (kJ)
1 J/s = 1 watt (W)

POTENTIAL ENERGY OF FOODSTUFFS

This is determined by measuring the heat produced when a sample is burned in the bomb calorimeter. Typical values are:
- A mixture of carbohydrate yields 17 kJ/g.
- Fat yields 39 kJ/g.
- Protein yields 17 kJ/g (after subtracting the potential energy of urea). 1 g of urinary nitrogen arises from the metabolism of 6.25 g of protein.

METHODS OF ESTIMATING ENERGY OUTPUT

Energy expenditure can be measured by:
- *Direct calorimetry* – measurement of heat production in a human calorimeter.
- *Indirect calorimetry* – measurement of oxygen utilisation.
  Energy expenditure can also be measured from the duration and type of physical activities during each 24-h period (keeping a diary), then looking up the calorific value from tables and finding the sum.

THE RELATIONSHIP BETWEEN OXYGEN CONSUMPTION AND HEAT PRODUCTION

There is a relationship between the oxidation of carbohydrate and heat produced.

$$C_6H_{12}O_6 \quad + \quad 6O_2 \quad = \quad 6CO_2 \quad + \quad 6H_2O \quad + \quad heat$$
$$180 \text{ g} \qquad 6 \times 22.4 \text{ L} \quad 6 \times 22.4 \text{ L} \qquad 18 \text{ g } 2.78 \text{ MJ}$$

$$1 \text{ g glucose} = \frac{2.78}{180} = 15.5 \text{ kJ (3.69 kcal)}$$

or 1 L of oxygen consumed produces 20.8 kJ (4.95 kcal)

$$\text{Respiratory quotient (RQ)} = \frac{\text{Vol. of } CO_2 \text{ produced}}{\text{Vol. of } O_2 \text{ utilised}}$$

$$RQ_{glucose} = \frac{6 \times 22.4}{6 \times 22.4} = 1$$

Similar equations can be written for protein and fat.

The energy equivalent of oxygen is approximately 20 kJ (4.8 kcal) for each litre of oxygen utilised.

The RQ determines the precise value of oxygen utilised.

Energy yields from oxidation of foodstuffs

Quantity (1 g)	O_2 required (ml)	CO_2 produced (ml)	RQ	Energy developed (kJ)	Energy equiv. of 1 L O_2
Starch	828.8	828.8	1.0	17.51	21.13
Animal fat	2019.2	1427.3	0.707	39.6	19.62
Protein	966.1	781.7	0.809	18.59	19.26

The RQ gives a clue to the type of food being consumed and oxidised
The RQ on a mixed diet is about 0.85.

Methods

Indirect calorimetry – closed circuit using the spirometer
The CO_2 produced is absorbed and the O_2 consumption is measured by the reduction in level of the spirometer bell.
The RQ is assumed to be 0.8 and the calorific value of oxygen to be 20.2 kJ/L.
Spirometer – open circuit
The subject breathes air into a Douglas bag.
The volume of air collected is measured and corrected to standard temperature and pressure (STP).
The O_2% is measured and the O_2 consumption calculated.
The CO_2 is measured and the output calculated.
The energy output is calculated after determining the calorific equivalent of O_2 at the RQ measured (from tables).
Weir's formula. This is a quick way of calculating energy expenditure by just collecting expired air.
The energy value of 1 L of expired air is:
$0.209 (O_1 - O_e)$ kJ where: O_1 = %O_2 in inspired air
O_e = %O_2 in expired air
The errors involved in not estimating CO_2 are ± 0.5% and urinary nitrogen are ± 1.0%.
Methods for use away from the laboratory:
- Max Planck respirometer.
- Integrating motor pneumotachograph (IMP).
- In clinical practice:
  Record of energy expenditure. An hour diary is kept and the different types and duration of physical activity recorded. The energy values for each activity are then obtained from tables. Doubly labelled water (D_2O^{18}) technique.

METABOLIC RATE

For an individual, the output of energy (and rate of oxygen consumption) depends upon:

- Body size, % of lean body mass (fat has a lower metabolic rate than lean body mass (LBM)).
- Muscular activity – accounts for 20–40% of total energy expenditure.
- Age – metabolic rate declines from birth to old age (decline in LBM and increase in fat).
- Nature and amount of food eaten:
  - postprandial thermogenesis (PPT)
  - diet-induced thermogenesis (DIT).
- Changes in body and environmental temperature:
  - fever
  - air temperature, wind speed, radiation from surroundings, cold-induced thermogenesis.
- Hormonal status, e.g. thyroid activity.
- Emotional status, e.g. excitement.

Basal metabolism

The basal metabolic rate (BMR) is the energy necessary to maintain essential processes, such as the beating of the heart, ventilation, the maintenance of body temperature, etc. The conditions for measurement are usually in the morning with the subject relaxed and warm in bed and 12–15 h after the last meal. By convention, the BMR is standardised to the surface of the body and is expressed as $kJ/m^2/h$.

Surface area is obtained from nomograms based on the Du Bois formula:

$$S = W^{0.425} \times H^{0.725} \times 0.007184$$

where:

S = surface area in m^2
W = nude weight in kg
H = height in cm.

The BMR varies with age and sex with an average of 167 $kJ/m^2/h$ for males and 150 $kJ/m^2/h$ for females. The difference between males and females disappears if the energy consumption is expressed in terms of LBM. Also in people whose body shape differs markedly from the normal, the use of surface area may mislead.

MEASUREMENT OF LEAN BODY MASS

See page 306.

POSTPRANDIAL THERMOGENESIS (PPT) OR SPECIFIC DYNAMIC ACTION (SDA) OF FOODS

Food stimulates metabolic activity and heat production, to an extent of about 10% of the energy content of the diet. This effect is called the PPT or SDA. The causes of PPT are:

- The mechanical act of eating.

- The work of the digestive glands and the digestion and transport of ingested nutrients.

MECHANICAL EFFICIENCY OF THE HUMAN BODY

The mechanical efficiency of the human body can be measured.

$$\text{Efficiency} = \frac{\text{useful work}}{\text{potential energy of fuel}} = \frac{\text{useful work}}{\text{useful work} + \text{heat}}$$

Measurement
External work is measured, e.g. on a bicycle ergometer.
O_2 consumption is measured, i.e. the potential energy of the fuel.
Gross efficiency 10–20%.

Energy expenditure
Food provides energy for three main purposes:
1. Basal bodily activities (BMR) – 7.7 MJ/day.
2. Everyday activities – sitting, standing, walking and dressing – 1.5–2.0 MJ/day.
3. Work – depending upon occupation – up to 1.6 MJ/h.

In computing the energy output from tables, the following must be taken into account:
- Energy expenditure varies from minute to minute in any job.
- In most types of occupation work will vary from heavy to sedentary.
- Very heavy work cannot be maintained for long.
- For some types of work, e.g. walking, both speed and gross body weight must be taken into account.

To assess activity accurately a detailed minute-to-minute diary must be kept for a complete week.

Classification of severity of work

Grade of work	kJ/min
Very light	< 10
Light	10–20
Moderate	20–30
Heavy	30–40
Very heavy	40–50
Exceedingly heavy	> 50

(Reproduced from: Davidson, Passmore and Brock Human nutrition and dietetics. Churchill Livingstone, Edinburgh)

HUNGER AND APPETITE

DEFINITIONS

Appetite – the desire for food.
Hunger – sensation of emptiness when abstaining from food.

BODY WEIGHT

The mechanisms involved in controlling body weight are poorly understood, but in most people weight is constant over a number of years and any changes are slow in the face of considerable variation in intake. It is usually stated that when:
Energy intake equals energy output the weight is constant.
Energy intake is greater than output then obesity will occur.
Energy intake is less than output weight loss will occur.

CONTROL MECHANISMS

The voluntary control of intake is determined by a balance between appetite and satiety, and this is probably achieved by both hormonal and neural influences. The control of body weight involves:
- Hypothalamic centres:
    A medial or satiety centre
    A lateral or feeding centre.
- Receptors for monitoring food intake:
    Oesophageal responding to distension
    Stomach also responding to distension.
- Neural pathways – afferent in the vagus nerves.
- There may be receptors monitoring blood glucose, fatty acids and ketone bodies.
- Role of CCK-PZ:
    Inhibition of gastric emptying
    A satiety effect.
- Glycogen stores – play a role in appetite, low stores have a stimulating effect.

THE NORMAL DIET

Daily requirements vary according to:
- Age
- Sex – pregnancy
- Physical activity
- State of health.

WHAT SHOULD WE EAT?

This depends upon appetite, personal preference, social habits, and the need to prevent disease. Dietary factors are implicated in the cause and prevention of disease.
    The diet should provide sufficient carbohydrate and fat to meet most of the energy expenditure, but with the proviso that fat should

be 30% or less of the energy intake and that saturated fats should be less than 10% of the calories.

When considering fats there are two types:
- Animal fats which are usually saturated:
  butter
  lard and suet.
- Fat from plants and fish (oils) – generally unsaturated.
  Fats influence diets by:
- Affecting palatability.
- Influencing satiety.
- Having a high calorific value (39.12 kJ/g = 9.3 kcal/g).

PROTEINS

Proteins are required in the diet for:
- Construction of new tissue in growing individuals.
- Building up new tissue after a wasting illness.
- Making good tissue wear and tear.
- Supplying amino acids for synthesis of hormones and enzymes.
  Protein turnover (breakdown and synthesis) in the adult is about 2.5 g/kg/day.

Food requirements
Food intake should be balanced against energetic requirements to maintain appropriate body weight and for growth and development, e.g. in childhood and pregnancy.

Protein requirements
Adults
For a normal adult an intake of 1 g/kg body weight/day is often recommended, though a range from 0.21 to 0.65 g/kg has been observed to maintain equilibrium. A safe figure would be 0.7 g/kg/day if the protein is of good quality.

Children
Children require food for normal and sustained growth. Growth is fastest in first 3 months of life – 2–3 g/kg/day, falling to about 1.25 g/kg/day declining until growth stops. Protein cannot be utilised for growth until energy needs are met.

ESSENTIAL FACTORS IN FOOD

Vitamins
For requirement see under individual vitamins.

Minerals
For calcium and potassium see page 322; for iodine see page 263.

Sodium
The intake of salt should be limited to 6 g/day or less. High intakes are associated with hypertension.

Essential fatty acids
Linoleic, linolenic and arachidonic acids are so called because they are required in small quantities for normal health but cannot be made within the body.
- Linoleic acid (has two double bonds), contained in large amounts in vegetable seed oils – maize, sunflower.
- Linolenic acid (has three double bonds), occurs in small quantities in vegetable oils, especially linseed oil.
- Arachidonic acid (has four double bonds), occurs in very small amounts in animal fats but can be made from linoleic acid in the body.

Essential amino acids
There are two groups of amino acids:
1. Those readily synthesised by the body.
2. Those which must be supplied in the diet, called essential amino acids.

The definition of an essential amino acid is one which is not synthesised in the body out of materials ordinarily available, at a speed commensurate with normal growth, e.g. lysine and threonine. In the hormone insulin, 57% of its constituent amino acids are essential amino acids.

Water and electrolytes
Water intake is determined by thirst (see Ch. 9) for water and electrolytes.

Nitrogen balance
If the amount of nitrogen in the diet equals the amount of nitrogen excreted the individual is said to be in nitrogen balance. For humans, about 30 g/day of first-class protein (see below) are required to maintain nitrogen balance.

Nitrogen output exceeds intake→negative balance (starvation and malnutrition; after surgery).

Nitrogen output less than intake→positive balance (during growth, tissue repair and pregnancy).

In the absence of protein in a diet that is otherwise adequate, the nitrogen output falls to 2–3 g/day:
- Tissue protein is broken down.
- Obligatory loss in the faeces is 1 g/day.
- Loss of skin, nails, hair and sweat is 1 g/day.
- During menstruation, loss is 2–3 g.

Dietary or nutritional value of a protein
This is dependent upon the nature and proportion of its constituent amino acids. It is measured as a biological value or a chemical score.

Biological value
A protein of high biological value supplies all essential amino acids in optimal proportions, and is able to maintain nitrogen balance when they are the sole source of protein, e.g. most animal protein.

A protein of low biological value is deficient in one or more essential amino acids, e.g. zein of maize is deficient in tryptophan and lysine.

Biological value is defined as:

$$\frac{\text{Dietary nitrogen retained}}{\text{Dietary nitrogen absorbed}} \times 100$$

when the protein under test is fed to a subject on an otherwise protein-free diet and when endogenous nitrogen output is constant.

Chemical score
This is the proportion of each essential amino acid in a foodstuff expressed as a percentage of the same amino acid in whole egg protein. The lowest percentage obtained is the chemical score. Protein requirements have been listed above, but 10–15% of the total energy should be in protein form with 60% of a high biological value, i.e. animal. For vegetarians a good mix of suitable vegetables should substitute.

Protein deficiency
The body does not possess a protein store. Inadequate protein intake results in:
- Tissue protein breakdown.
- Different organs are depleted at different rates.
- Liver > kidney > skeletal muscle.
- Malignant tumours grow.
- Plasma albumin is eventually reduced, giving rise to a low oncotic pressure and oedema.

Some organs may continue to grow and wounds heal.

Kwashiorkor – severe protein malnutrition
In the Ga language of Ghana *kwashiorkor* means 'the disease the first child gets when the next one is on the way'. This is a disease of early life, but children and young adults can get it.

The features of this disease are:
- Growth failure.
- Preservation of subcutaneous fat with wasting of muscles.
- Preservation of buccal pads of fat giving rise to a 'moon-face'.
- Oedema – correlates with the level of serum albumin.
- Apathetic, anorexic, miserable, whimpering cry.
- Desquamation – 'crazy-paving dermatosis' with depigmentation.

Effects of starvation
During severe starvation:
- Carbohydrate and fat stores run down.
- Glycogen is soon exhausted.
- Fat supplies energy for some weeks.
- Metabolising fat gives rise to ketosis.
- Protein breakdown begins before fat stores are exhausted.
- Amino acids are liberated to provide energy.
- Atrophy of muscle occurs.
- Nitrogen excretion is low until stores of carbohydrate and fat are exhausted.

THE VITAMINS

These substances are essential organic compounds required in metabolic processes. If deficient, a 'biochemical lesion' develops and structural and functional changes follow. Daily requirements are quite small and depend upon:
- Period of growth.
- Pregnancy and lactation.
  Vitamins are not usually synthesised by the body.

Vitamin A (retinol)
There are two forms, retinol and 3-dehydroretinol. The sources of vitamin A include:
- Plants – carotene (precursor of vitamin A), dark green leafy vegetables.
- Animals – vitamin A esters, in liver, fish liver oil, eggs, butter, cheese.

Physical and chemical nature
Lipid soluble.
Unsaturated alcohol ($C_{20}H_{30}O$) corresponding to half a molecule of beta-carotene.
Readily oxidised; slowly destroyed in oils and fats especially if rancid.
Frying food destroys vitamin A.
Cooking of vegetables reduces carotene only slightly.

Absorption
Vitamin A in the diet:
   Incorporated into bile salt micelles
   Hydrolysed by pancreatic esterases
   Absorbed and re-esterified in the mucosal cell
   Transported via the lymph and blood to the liver.
Carotene in diet:
   Incorporated into micelles
   Absorbed and converted in the mucosal cell into vitamin A
   Some absorbed intact.

Body stores – liver store lasts several months.
Transport in plasma – attached to albumin and lipoproteins.
Concentration in plasma in health:
  Vitamin A 200–500 µg/L
  Carotenoids 200–2000 µg/L.

Deficiency
Inadequate intake.
Defective absorption, e.g. in obstructive jaundice (absence of bile salts).
Xerophthalmia (cause of blindness in developing countries).
Retinal abnormalities, night blindness (lack of visual purple formed from retinol).
Keratinisation of cornea, scarring and keratomalacia, follicular hyperkeratosis.
Damage to respiratory epithelium.

Excessive intake of vitamin A
Drowsiness.
Headache.
Vomiting.
Skin changes and loss of hair.
Anaemia.

Recommended intake for adult
750 µg/day.

Thiamin (vitamin B$_1$)

Physical and chemical nature
Water soluble.
Pyrimidine and thiazol joined by a methylene bridge.

Sources
Plants, pulses, nuts, outer coat of grain.
Wheat flour a poor source unless artificially fortified.
Wholemeal flour 350–400 µg%.
Lean part of meat.
Yeast and yeast extract.

Blood content
5–11 µg% (as TPP).

Mode of action
Thiamine pyrophosphate (TPP) is a co-enzyme in decarboxylation of α-oxoacids, such as pyruvic acid.

Deficiency
Decreases glucose utilisation.

Prolonged deficiency leads to beri beri:
   Polyneuropathy
      Peripheral nerve degeneration
      Muscle weakness and atrophy
      Disturbances of sensation.
   Enlargement of heart
   Generalised vasodilatation
   Cardiac failure with oedema.

Recommended intake
1 mg/day.

Riboflavin (vitamin B$_2$)

Physical and chemical properties
Water soluble.
Riboflavin is converted in the cell to:
      Riboflavin phosphate
      Flavin adenine nucleotide
      Both combine with protein to give the flavoproteins which act
      as hydrogen carriers.

Stability
Relatively stable in acid solutions in the dark.
A little may be lost in cooking.
Considerable loss from milk in bright sunlight.

Sources
Synthesised by plants, germinating seeds, peas, beans and yeasts.
Animal sources – milk, egg yolk, liver, kidney, heart.
Absorbed from the upper part of the gut.

Blood content
100 mM/L (as FAD and riboflavin phosphate).

Deficiency
Skin becomes rough and scaly.
Lips bright red swollen and cracked.
Corners of mouth fissured.
Tongue tender and magenta coloured.
Cornea may become vascularised.

Recommended intake
1.5–1.8 mg/day.

Niacin (nicotinic acid)

Physical and chemical properties
Water soluble
Nicotinic acid is converted into active derivatives:

Nicotinamide adenine dinucleotide (NAD)
Nicotinamide dinucleotide phosphate (NADP).

Stability
Not destroyed by heat, light or alkali.

Sources
Plants – outer part of the grain, but is removed by milling, legumes, yeast.
Animals – liver, kidney, muscle, poultry.

Blood content
50 μmol/L.

Physiological role
Hydrogen transport.

Deficiency
Prolonged general ill health.
Irritability, depression.
Loss of weight, strength and appetite.
Diarrhoea.
Skin reddened, rough and scaly.
Redness and soreness of the mouth.
Swelling and redness of the tongue.

Requirements
12 mg/day.

Pyridoxine (vitamin B$_6$)

Physical and chemical properties
Water soluble and is involved in the metabolism of amino acids (and conversion of tryptophan into nicotinic acid). It is necessary for the formation of haemoglobin.
Deficiency is rare.

Sources
Meats, fish, eggs, whole cereals, brussel sprouts, potatoes and bananas.

Requirements
Related to protein content of diet.
Pregnant women and those taking the contraceptive pill may benefit from an increased intake.
Very high intakes are dangerous.

Pantothenic acid

Physical and chemical properties
Water soluble. It is necessary in the chemical reactions which release energy from fat and carbohydrate.

Sources
It is widespread in foods and therefore deficiencies are unlikely.

Biotin

Physical and chemical properties
Water soluble.
Essential in the metabolism of fat.

Sources
Made by bacteria in large intestine, therefore there is little need for any to be provided by diet.

Folic acid

Physical and chemical properties
Water soluble.
Contains three moieties: pteroyl, paraminobenzoic acid, glutamic acid.

Sources
Plant tissues – fresh green vegetables, yeast, mushrooms.
Animal tissues – liver and kidney.

Absorption
Polyglutamate broken down to monoglutamate and absorbed by mucosal cell and converted to 5-methyl tetrahydrofolic acid (methyl THF).

Physiological action
Methyl THF converted to THF in cells.
Acts as a coenzyme carrying one-carbon units.
Involved in synthesis of DNA, purines, serine and histamine.

Deficiency
Due to dietary lack or malabsorption.
Disturbances of DNA synthesis lead to defective erythropoiesis.
Megaloblastic anaemia.
Neural tube defect.
Recommended intake 200 μg folic acid/day.
Supplements (0.4 mg/day) given at conception and in pregnancy to prevent neural tube defects (spina bifida) – eat foods rich in folic acid.

Cobalamin (vitamin B$_{12}$)

Physical and chemical properties
Water soluble.
Molecule consists of a modified metalloporphyrin with cobalt in the centre linked to a nucleotide.
B$_{12}$ consists of a number of cobalamin compounds:
   5-Deoxyadenosyl cobalamin – main form in the tissues and diet
   Methylcobalamin – main form found in the plasma
   Hydroxycobalamin.

Sources
All forms of animal tissue particularly liver where the store may last 4 years.

Absorption
It is actively absorbed in the terminal ileum as a complex with a glycoprotein. The glycoprotein is secreted by the parietal cell into the stomach. In the absence of acid secretion, glycoprotein (Castle's intrinsic factor) is also not secreted.

Plasma levels
Methylcobalamin is bound to two carriers, transcobalamins I and II – 250–1000 ng/L.

Stores
Liver.

Physiological role
5-Adenysyl cobalamin is a hydrogen acceptor and an essential co-enzyme in DNA synthesis.

Deficiency
Rare except in the true vegetarian.
Lack of intrinsic factor occurs in:
   Atrophy of gastric parietal cells
   Gastrectomy – it is 1 or 2 years before the deficiency is revealed.
Diseases of the ileum preventing absorption of vitamin B$_{12}$ – glycoprotein complex.
Deficiency features.
A megaloblastic anaemia (pernicious anaemia).
Subacute degeneration of the spinal cord.
Peripheral nerve damage.

Recommended requirement
2 µg/day.

Vitamin C (ascorbic acid)

Chemical and physical properties
Water soluble.

A powerful reducing agent.
Ascorbic acid ($C_6H_8O_6$).

Sources
Fruits and vegetables – citrus, guavas, turnip, broccoli greens and brussel sprouts are especially rich in the vitamin.
Animal – kidney, liver and roe.
High concentrations are found in:
    Adrenal gland
    Pituitary gland
    Corpus luteum
    Thymus.

Blood content
30–100 μM/L.
At over 60 μM/L vitamin C appears in the urine.

Deficiency
Defective formation of collagen.
Wound healing retarded.
Formation of bone abnormal as the result of the collagen defect.
Scurvy:
    Swollen bleeding gums
    Bleeding into skin and deeper tissues – increased capillary fragility
    Deficiencies in intercellular cement.

Recommended intake
30 mg/day.

Vitamin D (calciferol)

Physical and chemical properties
Fat soluble.
The name vitamin D is applied to a number of compounds:
    Vitamin D_2 – ergocalciferol
    Vitamin D_3 – cholecalciferol.
Vitamin D_2 is produced by ultraviolet irradiation of ergosterol (a plant sterol).
Vitamin D_3 is produced by irradiation of 7-dehydrocholesterol in the skin.
Vitamin D is heat stable, only small losses occur in storage, cooking, canning or drying food.

Sources
Fish liver is the richest source. Fatty fish, salmon, herrings, and sardines are good sources.

Absorption
In the upper half of the small intestine in the presence of bile salts.

Storage and carriage
Stored in adipose tissue and muscle.
There is sufficient to postpone deficiency symptoms for years.
Bound to plasma α-globulin.

Physiological actions
Vitamins D_2 and D_3 are hydroxylated in the liver, forming
25-hydroxycholecalciferol. 25-Hydroxycholecalciferol is the most
important form of vitamin D in plasma. Further hydroxylation
occurs in the kidney to form 1,25-dihydroxycholecalciferol
(1,25-DHCC – calcitrol) and 24,25-dihydroxycholecalciferol (24,25-
DHCC). 1,25-DHCC acts as a hormone:
    Production of 1,25-DHCC by the renal tubule is controlled by
    plasma parathyrin and hence by plasma calcium; therefore,
    output is increased by a low plasma calcium concentration when
    the plasma calcium is high; the principal metabolite is 24,25-
    DHCC.

Action of 1,25-DHCC
On the small intestine, promoting active absorption of calcium
Binds to receptors in mucosal cell nucleus stimulating the synthesis
of calcium binding protein.
Acts also on:
    Bone to stimulate activity of osteoclasts
    Kidney proximal tubules
    Parathyroid gland
    Muscles.

Deficiency
Two causes, lack of vitamin D in the diet and lack of response to
ultraviolet radiation.
In the UK vitamin D deficiency is particularly found in Asian
immigrants and in house-bound elderly women.
Defects in absorption caused in:
    Coeliac disease (epithelium defective)
    Obstructive jaundice (lack of bile salts).
Deficiency causes inadequate deposition of calcium salts in newly
formed bone matrix:
    Rickets in children
    Osteomalacia in adults.

Recommended intake
Infants and young children 10–18 µg/day.
Adults and older children 2.5 µg/day.

Vitamin E

Physical and chemical properties
Fat soluble.

A series of chemically related compounds.
α-Tocopherol the most important. A biological antioxidant that protects from oxidation:
 Polyunsaturated fatty acids
 Vitamin A and carotene
 Protects membrane lipids against peroxidation.

Sources
Found in greatest concentration in vegetable oils.

Physiological function
Not known with certainty, but it is thought that along with vitamin C it mops up active radicals which may be involved in the pathogenesis of coronary heart disease and malignancies.
 Some margarines are now fortified with vitamin E.

Vitamin K

Physical and chemical properties
Fat soluble.
Two forms:
 Vitamin K_1 phylloquinone
 Vitamin K_2 menaquinone.

Sources
Vitamin K_1 – green parts of plants.
Vitamin K_2 – synthesised by intestinal bacteria.

Absorption
Depends upon presence of bile salts.

Storage
Very little is stored, depletion occurs rapidly.

Deficiency
Deficiency symptoms not produced in healthy subjects solely as a result of dietary lack.
Vitamin K is synthesised by intestinal bacteria, but deficiency can occur after surgery when taking little food, and when intestinal flora are suppressed by antibiotics.
Where there is malabsorption of fat.
Clotting impaired.
Reduction in prothrombin, and factors VII, IX and X.

MINERALS

For the metabolism of Ca and P see page 324, for Fe see page 4 and for I see page 263.

Sodium and chloride

All body fluids contain sodium chloride, especially the ECF. These elements are involved in maintaining the water balance of the body. Sodium is essential for nerve and muscle activity and the function of other organs. Too low an intake (or excessive loss) results in cramps and too much is associated with hypertension.

Absorption and secretion
These are dealt with elsewhere in this book (see p. 139).

Requirements
In a temperate climate an adult needs less than 3 g a day and should not exceed 6 g a day.

Sources
Salt is relatively low in most foods, but is added to many prepared foods.

Magnesium

The total body content of Mg is 500 mmol with 60% in bone. It is found in cells and mostly bound to protein. It has been claimed that dietary deficiency may be involved in the aetiology of some cardiovascular disorders and vascular contractility and may modulate atherogenesis.

Plasma level
1 mM/L with 30% bound; 0.6 mM/L free.

Balance
Diet 20 mM/day: 8mM (40%) absorbed from gut.
About 8 mM secreted in urine/day.
Physiological function:
   As a component of most enzymes
   As a cofactor in reactions involving ATP
     in the stability of some intracellular organelles
     in the normal function of nerve and muscle.

Sources
Green vegetables and cereals.

Deficiency
There is never a pure Mg loss, it is always associated with other intracellular ions – K and P.
Occurs during following conditions:
   Prolonged gastrointestinal aspiration
   Chronic watery diarrhoea
   Steatorrhoea (loss of fat in faeces)
   Chronic alcoholism
   Chronic administration of diuretics.

Consequences:
Anorexia
Cell atrophy
Negative nitrogen balance
Cardiac arrhythmia and cardiac arrest.

Copper
It is an essential factor involved as a component of cytochrome oxidase and other enzymes.

Transport
Bound to caeruloplasmin.

Sources
Liver, shellfish, meat, bread, cereal products and vegetables.
Excess gives rise to hepatolenticular degeneration.

Deficiency
Not known in adults but can occur in malnourished infants, especially if fed on cows' milk alone as it is very low in copper.

Zinc
An essential factor involved:
As a component of many enzymes e.g. carbonic anhydrase.
In non-enzymatic free radical reactions.
Helps in the healing of wounds.

Sources
Herring, beef, liver, eggs, cheese and nuts.

Deficiency
May be brought about by excess phytic acid in the diet preventing absorption.
Growth and sexual development retarded.
Atrophy of the thymus.

Manganese
Essential factor involved:
As a component in many enzymes.
In the function of mitochondria.
As a cofactor in some enzymatic reactions.

Sources
Tea, nuts, spices and whole cereals and plant products.

Deficiency
No recognised disease.

Cobalt
A component of cobalamins (see above).

Deficiency
Deficiency may lead to anaemia.
Excess gives rise to toxic effects.

Fluoride
Plays an important role in the stability of hydroxyapatite crystals and has an anticaries function.

Optimal intake
1–2 mg/day.

DIET AND HEALTH

Epidemiological studies have shown certain diets are associated with cancer, coronary heart disease (CHD), birth defects and cataracts.

DIETARY FAT AND CORONARY HEART DISEASE
A high intake of:
- Saturated fat leads to a high incidence of CHD.
- Unsaturated fat has a low incidence of CHD.

In this respect it is recommended that at least two portions of fish, one of which is oily fish should be eaten weekly. Fat-reduced dairy products and spreads should be used instead of full fat products.

	Butter, margarines and low-fat spreads		
	Saturates (g/100 g)	Trans isomer (g/100 g)	Saturates + trans (g/100 g)
Butter	54.8	3.4	58.2
Margarine, hard animal and vegetable fats	34.6	12.2	46.8
Margarine soft (not polyunsaturated)	27.2	9.8	37.0
Margarine soft (polyunsaturated)	17.0	6.7	23.7
40% fat (not polyunsaturated) Spread	11.7	4.1	15.8
40% fat (polyunsaturated) Spread	8.9	0.7	9.6
20–25% fat spread	6.2	2.3	8.5

(Figures from the British Heart Foundation.)

Although the role of trans fatty acids in the aetiology of CHD remains unresolved, it appears prudent to reduce the intake of both trans and saturated fatty acids, both of which may be atherogenic.

OTHER FACTORS

Diets rich in vegetables and fruit have been shown to reduce the risk of various cancers and CHD. The factors responsible are unknown; the antioxidation properties of vitamins C, A and E have been suggested, but such claims are as yet unsubstantiated.

Care must be taken in interpreting epidemiological data in the quest for a perfect diet, for a population with a low incidence of CHD has a high incidence of gastric cancer and haemorrhagic stroke (Japan).

CHOLESTEROL

Serum cholesterol levels predict CHD and these levels increase with increasing intake of fat and this is the reason for the above recommendations. There are populations – e.g. Greek – which have high fat intakes and a low incidence of CHD, but the content is low in saturated fat but high in unsaturates, e.g. olive oil. These populations also have a high intake of vegetables. Reducing the levels of cholesterol in the blood does reduce the incidence of CHD.

6. The nervous system

THE NEURONE

Neurones are found in the brain, spinal cord and autonomic ganglia but as many neurones are present in the nerve plexuses in the gastrointestinal tract. The neurone is a highly specialised cell that functions to receive, assimilate and pass on information.

In order to process information nerve cells have developed specialised structures. Thus, the component parts of the neurone and the connections between neurones, the synapses, have specific functions in this information processing.

THE CELL BODY

Every neurone has a cell body. This is the central part of the cell and contains the cell nucleus.

DENDRITES

Fine processes emanate from the cell body which can divide into finer processes. These structures are called the dendrites. The network of fine branches of the dendrites is called the dendritic tree. The dendritic tree greatly increases the area of the cell body available for contact with other nerve cells. The size and extent of the dendritic tree can vary in different areas of the brain, with some large cells having as many as 10 000 synapses. The structure of the dendritic tree reflects the function on the neurone to gather information. The dendrites are the area of the neurone where electrical stimuli are received and integrated.

EXCITATORY AND INHIBITORY SYNAPSES (Fig. 6.1)

The synapses, or terminal boutons, release transmitter substances when an action potential invades the synapse (see p. 181). Transmitter substances can be either excitatory or inhibitory on the postsynaptic membrane, giving rise to a small depolarisation. This potential change is small, often less than 0.5 mV, and it moves the membrane potential closer to threshold for the initiation of an action potential. It is therefore excitatory and is called an *excitatory postsynaptic potential* or EPSP. Inhibitory synapses give rise to a

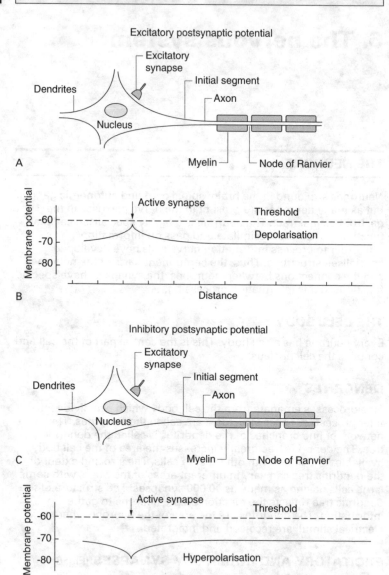

Excitatory postsynaptic potential

A

Inhibitory postsynaptic potential

C

Fig. 6.1
Summary of electric events at an activated excitatory synapse (A, B) and an inhibitory synapse (C, D). A, Spatial distribution of the depolarising synaptic potential. B, Diagrammatic anatomy of the nerve-cell body corresponding to A. C, Spatial distribution of the hyperpolarising synaptic potential. D, Diagrammatic anatomy of the nerve-cell body corresponding to A.

small hyperpolarisation of the postsynaptic membrane of approximately 0.5 mV. This shift in membrane potential is further away from threshold and reduces the likelihood of an action potential being fired. This process is called an *inhibitory postsynaptic potential* or IPSP. The EPSPs and IPSPs are transient events lasting 2–5 ms.

Ionic events   The transmitter substances act at specific molecules on the postsynaptic membrane – *receptors*. The receptors are linked to ion channels which are opened when the receptor is occupied.

Excitatory synapses
Excitatory synapses, activated by excitatory transmitters such as acetylcholine, serotonin and glutamate, are linked to cation channels which when open conduct predominantly Na^+ ions. Na^+ moves into the cell, resulting in a depolarisation of the membrane.

Inhibitory synapses
Inhibitory synapses, activated by transmitters such as glycine and γ-amino-butyric acid (GABA), are linked to anion channels carrying Cl^- ions when they are open. Since the Cl^- equilibrium potential (see p. xx) is more negative than the resting potential, Cl^- enters the cell, giving rise to a hyperpolarisation of the neurone membrane.

EPSPs and IPSPs can add up to produce a net synaptic potential. If the balance of synaptic input is excitatory, then the membrane potential of the neurones may reach threshold and fire an action potential. If the input is predominantly inhibitory, the membrane will not reach threshold and the nerve will not fire.

Temporal summation
Synapses on the cell body may be activated in quick succession. When this happens the EPSPs or IPSPs will add together. This is called temporal summation.

Spatial summation
Synaptic potentials generated in adjacent regions of the dendritic tree will add together if they are activated close together in time. Both EPSPs and IPSPs can be summed in this way, a process called spatial summation.

The dendrites of many neurones are not capable of initiating an action potential. Action potentials are generated at a specialised region of the cell body, close to the origin of the axon, called the *axon hillock*. The EPSPs and IPSPs are local potentials which are largest at the site of the synapse. In adjacent regions of the dendrite the potential gets progressively smaller (decremental conduction). For this reason, EPSPs and IPSPs in the periphery of the dendritic tree are not as effective as synapses close to the axon hillock. This is an important aspect of spatial summation, as the site of a synapse on the dendritic tree influences the weighting of the information arriving on the neurone, influencing the likelihood of it firing.

AXON

The axon is a single long process which leaves the cell body and conducts electrical signals away from the cell body. An axon can travel very large distances. The motor axons leaving the spinal cord travelling to the foot may be 1 m or longer in length. At the periphery axons often branch extensively, giving rise to many synaptic terminals which can contact several cells.

The cell body at the site of origin of the axon is the axon hillock. It is in this region that the integration of information at excitatory and inhibitory synapses can result in depolarisation to threshold, resulting in the generation of an action potential. Once initiated, the action potential travels down the axon to the target site of the neurone.

SCHWANN CELLS

Axons of many neurones are surrounded by a specialised cell type, the Schwann cell. These cells wrap themselves around the axon forming layers of membrane, the myelin sheath. This sheath functions to electrically insulate the axon membrane. Regular gaps appear in the myelin sheath, *nodes of Ranvier*, where the axon membrane is relatively exposed. There is a high concentration of Na^+ channels at the nodal regions and this is where the action potentials form during nerve conduction. The inputs jump from one node to the next – saltatory conduction. This process is faster than the direct spread of the action potential along the uncovered cell membrane and greatly speeds the conduction of the nerve impulse along the axon. Saltatory conduction is also more energy efficient for the neurone.

THE SYNAPSE

The nerve–nerve synapse has many similarities to the nerve–muscle synapse or neuromuscular junction.

Structure
The neural synapse is made up of the bulbous terminal, a synaptic cleft separating the presynaptic terminal from the postsynaptic membrane. The nerve terminal or terminal bouton contains mitochondria, Golgi apparatus and vesicles (plain and coated). The vesicles contain transmitter substance and when the terminal is activated they migrate to the presynaptic membrane. The vesicles then attach to the membrane and fuse with it – a process known as exocytosis. As a result of this process, the transmitter substance is

released towards the target cell. Once the vesicle has fused and its membrane incorporated into the presynaptic membrane, a second sequence of events is triggered which retrieves this membrane – endocytosis. The early events involve a protein, *clathrin*, which is associated with areas of presynaptic membrane. This coated area is pulled into the cell forming pits. These indentations get deeper until a spherical bud is formed – the clathrin-coated vesicle. This structure is then modified and the coat is removed. The vesicle then fuses with the endosomes in the terminal from where new vesicles are formed, ready for filling with transmitter and release.

Function
On arrival of a presynaptic nerve impulse there is an influx of Ca^{2+}. This facilitates the movement of vesicles to the presynaptic membrane and the subsequent fusion of the vesicles. The vesicular contents are then passed into the synaptic cleft to diffuse to the postsynaptic membrane. The postsynaptic membrane has receptors for the transmitter substance. When these receptors are occupied the ion channels open and ions flow. Depending on the type of ion channel anions or cations will flow leading to excitation or inhibition respectively.

GLIAL CELLS (Fig. 6.2)

Glial cells are non-excitable cells which are closely associated with nerve cells both in the brain and in the periphery. One of their major

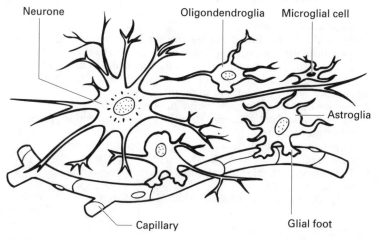

Fig. 6.2
Glial cells. (Reproduced with permission from: Rhoades and Pflanzer, WB Saunders Co. Ltd)

functions is to surround nerve processes with their membranes providing a layer of electrical insulation, *myelin*. This insulation is important in the rapid conduction of nerve impulses along the nerve axons. In the central nervous system glial cells outnumber nerve cells typically 10 : 1. Glial cells, unlike nerve cells, retain their ability to divide and proliferate in the adult brain. This can occur when areas of the brain are damaged. Glial cells, surrounding the damaged area, proliferate forming a glial scar. Specialised glial cells, microglia (see below), act as phagocytes to remove debris from the damaged area.

GLIAL CELLS IN THE PERIPHERY

In the peripheral nervous system the glial cells, or *Schwann cells*, are found wrapped around the axons forming a tight spiral of membranes. This sheath electrically insulates the nerve axonal membrane. It is interrupted at regular intervals, by *nodes of Ranvier*, which occur every 2 mm, and it is in this region that the nerve impulses occur. An action potential at one node depolarises the adjacent node by passive spread of current along the insulated region of the axon. This depolarisation then initiates an action potential at the second node. In this way the nerve impulse jumps from node to node along the nerve, a process termed saltatory conduction (Latin: *saltare* = to jump).

GLIAL CELLS IN THE CENTRAL NERVOUS SYSTEM

There are several types of glial cell in the central nervous system:

Oligodendrocytes
These glial cells are found mainly amongst the myelinated central axons where they are involved in the production of the spiral myelin sheath. They typically have a dark cytoplasm and nucleus, heavily condensed chromatin and few neurofilaments. A number of functions have been ascribed to them:
1. To take up and store transmitter substances.
2. To serve as a spatial barrier to isolate neurones and regulate ionic interactions.
3. To provide a structural framework to support the nerve networks.
4. To store and transfer metabolites between capillaries and neurones.
5. To guide the growth of axons during development.

Microglia
The microglia are small ovoid cells capable of acting as phagocytes. These cells are important in removing damaged cells at a site of injury.

Astrocytes

Astrocytes are star-shaped cells with long processes radiating from the cell body. They are usually larger than oligodendrocytes and microglia. They are commonly subdivided into two classes, protoplasmic and fibrous. The fibrous astrocytes contain many filaments and are found extensively in the areas of the central nervous system containing myelinated axons, the white matter. Protoplasmic astrocytes contain few filaments and are predominantly associated with the cell bodies, dendrites and synapses of neurones. They are therefore found in the grey matter of the brain. Both fibrous and protoplasmic astrocytes are also in close apposition to capillaries and appear to link nerve cells to the blood vessels. For this reason it has been suggested that they may be involved in transferring nutrients from blood to nerve cells.

THE CENTRAL NERVOUS SYSTEM

SPINAL CORD

The spinal cord is a cylindrical structure slightly flattened dorsoventrally. It extends from the junction with the medulla at the foramen magnum C1 to lumbar segment L1. It is located in the spinal column which protects it. It is covered by the meninges and a cushion of cerebrospinal fluid (CSF). The cord is covered by:

- Pia matter – innermost layer adheres to surface of cord.
- Dura mater – outermost layer, a tube continuous with dura mater of brain at foramen magnum extending to the second sacral vertebra.
- Arachnoid mater – delicate membrane attached to inner surface of dura.
- Subarachnoid space – between pia and arachnoid is filled with CSF.

The spinal cord is segmented, with 31 pairs of spinal nerves attached to the cord by a dorsal sensory root and a ventral motor root. It is enlarged at the origin of the nerves to the limbs, i.e. C4 to T1 and L2 to S3. The cord has a central grey area shaped in a letter H the limbs of which are laminated. There are 10 layers of neurones, which receive fibres in an ordered sequence. The anterior horn of the H gives rise to α-motor neurones and to δ-motor neurones. From T1 to T12 and L1 to L2 is the intermediolateral cell column which gives rise to the preganglionic sympathetic fibres. Segments S2 to S4 give rise to the pelvic outflow of preganglionic parasympathetic fibres.

THE BRAIN

The brain has three major subdivisions:

- The brain stem.
- The cerebellum.
- The diencephalon and cerebrum.

THE BRAIN STEM

The brain stem consists of:

The medulla oblongata
Continuous with the spinal cord.
Contains discrete nuclei (grey matter).
Sensory and motor nuclei for mouth, throat and neck.
Groups of neurones involved in the control of:
  Cardiovascular system
  Respiration
  Movement and posture (vestibular nuclei)
  Secretory and motor function of the gut
  Groups of neurones passing to the spinal cord and to higher centres.

The pons
Situated between medulla and midbrain.
Contains the same ascending and descending tracts of fibres.
Contains groups of nuclei involved in the control of:
  Respiration
  Motor and sensory functions of the face.
Acts as a large relay station – cerebral cortex → pontine nuclei → cerebellar hemisphere.

The midbrain
Continuous with pons below and diencephalon above.
Dorsal surface is characterised by the presence of four rounded elevations:
  Superior colliculi – concerned with the processing of visual information
  Inferior colliculi – concerned with the processing of auditory information.
Lateral surface is formed by the cerebral pedunculi formed by:
  Fibres of the pyramidal motor system
  Corticopontine fibres.
It contains:
  The red nucleus
  Substantia nigra
  Mesencephalic reticular formation.

THE CEREBELLUM (a coordinating system for posture and motor movements)

The cerebellum can be seen on the dorsal aspect of the brain stem attached to it by three large bundles of fibres, the superior, middle and inferior cerebellar peduncles. It appears as a rolled-up structure – like a hedgehog nose to tail. In the midline is the vermis which is flanked on each side by the cerebellar hemispheres. The body of the cerebellum is subdivided by a fissure, the primary fissure, into anterior and posterior lobes. The posterolateral fissure separates the flocculonodular node (the 'tail') from the body of the cerebellum. A layer of grey matter covers the surface which has a large number of narrow folds, the folia. A section through the organ shows a central core of white matter and four pairs of cerebellar nuclei:

- Fastigial
- Globose
- Emboliform
- Dentate.

DIENCEPHALON

The thalamus (see p. 199)

The hypothalamus

The hypothalamus lies ventral to the thalamus and on the floor of the third ventricle. It contains nuclei which are responsible for internal homeostasis by regulating autonomic functions such as:

- Body temperature.
- Heart rate and blood pressure.
- Release of hormones by the pituitary gland.
- The expression of emotions.
- Regulation of food and water intake.

The cerebrum

The cerebrum is divided into two hemispheres by a longitudinal fissure, at the base of which is a thick band of white matter, the corpus collosum, which joins the two.

The surface of each hemisphere is enlarged by folding. The summit of each fold is a gyrus and the space between a sulcus. A deep sulcus is called a fissure. The outer layer is made up of layers of cells (grey matter with a rich blood supply).

The hemispheres are divided by the fissures into a number of named parts. There are two important named fissures or sulci – lateral and central.

The lateral fissure or sulcus separates off the temporal lobe from the parietal lobe and the frontal lobe above. The temporal lobe contains the primary auditory cortex. The central sulcus runs from the longitudinal fissure, in the middle of the hemisphere, to the lateral sulcus below. It separates the frontal and parietal lobes. In front of it lies the precentral gyrus, the primary motor area and behind the postcentral gyrus, the primary somatosensory area.

THE BASAL GANGLIA

The structures under this heading are important in motor disturbances called dyskinesias, characterised by involuntary purposeless movements. The basal ganglia lie between the thalamus and cerebral cortex in each hemisphere. These consist of:
• Claustrum
• Putamen
• Caudate nucleus
• Globus pallidus
• Subthalamic nucleus
• Substantia nigra.
Alternative names used:
Neostriatatum = striatum = putamen + caudate nucleus
Paleostriatum = palladium = globus pallidus
Lentiform nucleus = globus pallidus + putamen.

THE LIMBIC SYSTEM

This system is composed of a number of structures:
• Limbic lobe of cerebral cortex
    Cingulate gyrus
    Hippocampal gyrus
• Hippocampus
• Hypothalamus
• Amygdaloid nucleus.
    It is important in the regulation of behaviour (emotion, rage, pleasure, etc.) and in memory.

RECEPTORS

Receptors are separate cells or specialised regions of a cell which detect physical changes in the environment (Table 6.1). They can be classified with respect to their position in the body:

Exteroceptor:
Superficially in skin:
    Touch
    Pain
    Temperature.

Proprioceptor:
In muscles, tendons and joints:
    Stretch
    Tension
    Movement.

Interoceptor:
In or associated with the viscera

Pain
Tension
Chemicals.

Teloceptors:
Recognise events in the remote environment:
Sight
Smell
Hearing.

Receptors can also be classified by their function:
Mechanoreceptors
Thermoreceptors
Chemoreceptors
Nociceptors
Photoreceptors.
The anatomical and physiological basis of sensation is detailed in Table 6.1.

General properties of receptors
Receptors are sensitive to one type of stimulus or energy. Activation of the nerve produces the particular sensation in the brain. The area of the body stimulated is determined by the brain by knowing which nerves are active. This area is known as the receptive field. The information transmitted is determined by:

Properties of receptors
Sensitivity of the receptor
Intensity and nature of the stimulus – threshold
Adaptation – slowly or rapidly adapting.

Temporal factors
Frequency and rhythm of discharge
Receptor adaptation
Duration of stimulus.

Spatial factors
Area stimulated
Number of receptors activated.

Local factors
Excitability may be controlled by:
CNS (in some cases)
Inflammation (e.g. in skin and viscera).

CUTANEOUS SENSES

Four modalities of sensation are usually described:

Table 6.1 Anatomophysiological basis of cutaneous sensation

Receptors	Location	Receptive field	Effective stimulus	Sensory function
Tactile				
Rapidly adapting				
Pacinian corpuscle	Subcutaneous	100 mm^2	Vibration (400–600 Hz) movement	Vibration
Krause end bulb	Dermis of glabrous skin	2 mm^2	Vibration (10–200 Hz) movement	Touch Spacial analysis, intensity
Meissner corpuscle	Dermis of glabrous skin	12 mm^2	Low frequency vibration (5–200 Hz) movement	Flutter Spacial analysis, intensity
Hair follicles	Hair follicle	1.5 cm^2	Hair movement vibration (5–40 Hz)	Flutter Spacial analysis, intensity
Slowly adapting				
Merkel receptor	Base of epidermis	11 mm^2	Indentation and pressure	Spacial analysis, intensity
Ruffini ending	Dermis	60 mm^2	Stretch of skin, pressure	(Pressure touch)
C-mechanoreceptor	Dermo-epidermal boundary	2 mm^2	Indentation, slow movement	(Itch)
Thermoreceptors				
Cold	Base of epidermis	1 mm^2	Steady and falling temps. 40–10°C	Cold
Warm	Base of epidermis	1 mm^2	Steady and rising temps. 35–50°C	Warmth and heat
Nociceptors				
Mechanical?	Skin	3 mm^2	Pin prick, squeezing	Pain (sharp 1st)
Thermomechanical?	Skin	3 mm^2	Temp. 42°C severe mechanical algogenic chemicals	Pain (dull 2nd)

- Touch
- Warmth
- Cold
- Pain.

SENSE OF TOUCH

By the use of touch it is possible to distinguish various states, e.g. softness, hardness, wetness, sharpness, etc. The threshold to touch is measured using calibrated von Frey hairs. Touch sensation is punctate in distribution. An adequate stimulus is deformation rather than pressure. Flutter vibration results from repetitive mechanical stimuli. Temporal patterns of firing of the mechanoreceptors of the skin are necessary to appreciate textures, inequalities, sharpness, etc.

Distribution of tactile sensibility
Sensory nerves are generally associated with hair follicles where a nerve network exists around each hair root. Stimulus is movement of hair, which results in the deformation of nerve endings. This in turn leads to excitation of nerve fibres. Removal of hairs causes a reduction in tactile sensibility.

Receptive field
Large.

Adaptation
Is rapid and of two types:
  Peripheral
  Central.

Localisation determined by
Position at which the afferent nerve fibre enters the CNS.
Higher central connections.
Experience (learning).

Structure of nerve endings
These are either:
- Free naked nerve endings. Sites: abundant in hairy skin, innervate follicles, also in glabrous skin, deep fascia, and viscera → connected to C and Aδ fibres.
- Encapsulated nerve endings. Sites: mainly hairless or glabrous skin (skin of hands and soles of the feet, the lips, eyelids, mucosal surfaces, viscera). These are connected to Aβ fibres.

Examples of encapsulated endings
Mainly pressure and vibration sensitive
Pacinian corpuscles
Meissner's corpuscles
Merkel's disks.

PAIN

Pain is a perceptual experience but there is no agreed definition of pain. The intensity and quality of pain are affected by emotional considerations:
- Fear.
- Anxiety.
- Previous experiences, and it is influenced by attention, anticipation and suggestion.
- Psychological attitudes – pain is not experienced by some, even after serious wounding.

Subjective experiences
Pain produced by a maintained stimulus only varies slightly in intensity, whereas other sensations weaken or fade. The pain system can be so excitable that even the merest touch elicits an unbearable lightning pain. Long-lasting chronic pain has a destructive effect on the personality.

Measurement of pain
Several procedures have been devised to measure pain:
- Focused radiant heat on the skin.
- Controlled temperature pulses with lasers.
- Calibrated pressure on the Achilles tendon.
- Hypertonic saline injections into muscles.
- Ischaemic pain.
- Clinical test for pain is pin prick.

Pain receptors
Pain receptors are free nerve endings. This is assumed to be the case because only free endings are found in the cornea where the only sensation evoked by stimulation is pain. This does not mean that all free nerve endings subserve pain.

Pain-producing substances
Many chemical substances can cause pain:
- Bradykinin may participate in the pain of:
  Peritonitis
  Pancreatitis
  Angina pectoris – ischaemic pain
  Intermittent claudication – ischaemic pain.
- Potassium – released into damaged tissues.
- Changes in pH.
- Inflammation can lower the threshold to pain and can release pain mediators:
  Bradykinin
  Substance P
  Acetylcholine
  5-Hydroxytryptamine
  Histamine.

Pain nerves
It had been observed that pain evoked with a brief sharp stimulus to the foot is perceived as coming in two waves, distinct, sharp and brief, followed by an interval with no pain and a second wave which is long lasting, diffuse and aching. Experiments on humans have shown that the pain is mediated by two fibre groups:
1. Aδ fibres mediate a pain which is distinct, sharp and well localised.
2. C fibres mediate a pain which is unpleasant (burning) and diffuse.

Asphyxia blocks Aδ fibres and the first pain before C fibres. Local anaesthetics block C fibres and the second pain before Aδ fibres. Two types of pain are not experienced on the face because of the small conducting distance.

Pathways
See p. 200.

Classification of pain
Pain is usually divided into three types:

Superficial pain
Well localised.
Quality – sharp, pricking, stabbing, burning, short lived or enduring. Threshold lowered by inflammation.

Deep somatic pain
Generally diffuse, not well localised, often referred to a place distant from the source. Pain from neck muscles may be felt as an ache at the back of the head.

Visceral pain
Most viscera are insensitive to stimuli which would cause intense pain if applied to the skin. In the unanaesthetised person the viscera are:
• Insensitive to cutting or burning.
• Sensitive to stretching.
• Sensitive to other stimuli if the mucosa is inflamed (hyperaemic and oedematous).

Excitation of tension receptors sufficient to cause pain inhibits gastrointestinal motility. The pain is:
• Diffuse.
• Difficult to localise.
• Varies in intensity from mild to severe.

The pain is usually referred from:
• The colon – to the hypogastrium.
• The small intestine – to around the umbilicus.
• The stomach – to epigastrium.

Referred pain

Pain in a diseased organ is projected to a definite position on the surface of the body.

Organ	Site to which pain is referred
Heart (angina pectoris)	Left arm and precordium
Liver and gall bladder	Right upper abdominal quadrant, right scapula
Stomach	Epigastrium
Small intestine	Umbilicus
Colon – ascending	Suprapubic region
– sigmoid and rectum	Deep pelvis and anus
Kidney	Loin and groin
Bladder – fundus	Suprapubic region
– neck	Perineum and penis
Uterus – fundus	Suprapubic region
– cervix	Lower back and perineum
Testes – vas deferens	Pelvis and perineum
Seminal vesicles, prostate	Pelvis and perineum
Diaphragm	Tip of shoulder and neck

When the parietal peritoneum is irritated in visceral disease, pain is located at the site of irritation. There is tenderness and sometimes spasm of the local muscles – visceromotor reflex. Pain from the appendix is felt:
• Early – in the midline near the umbilicus.
• Later – in the right iliac fossa when the parietal peritoneum is involved.

Figure 6.3 shows the proposed neural connections of visceral referred pain.

Pain modulation

There are two physiological mechanisms by which pain can be controlled:
1. A peripheral afferent input system.
2. A central descending system.

Both systems seem to act at a common site. This is associated with cells of the substantia gelatinosa (SG).

The peripheral spinal gate control theory

A theory of antagonism between large cutaneous afferents and small pain fibres was based on the observation that counter-irritants, heat and massage, etc. will alleviate pain.

Impulses in the larger fibres inhibit cells of the SG, thus shutting 'the gate' to the ascent of information from the smaller pain fibres. Although the mechanism may not be in the original form as proposed by Melzack and Wall, it has led to several new techniques for combating pain.

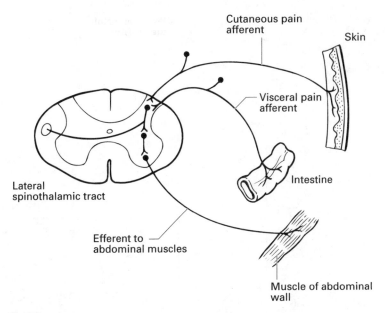

Fig. 6.3
Proposed neural connections of visceral referred pain.

The central descending system
Analgesia can be produced by electrical stimulation of the periaqueductal grey (PAG) matter and from parts of the limbic system. Descending fibres lie in the dorsolateral funiculus of the spinal cord, where control is exerted selectively on the pain input, e.g. in lamina V neurones transmission of large fibre information is unaffected but transmission of pain input is inhibited. The possible neural connections are seen in Figure 6.4.

e.g. PAG → nucleus raphe magnus → dorsolateral funiculus → substantia gelatinosa and laminae V, VI and VII

Some of the descending control of pain may be due to the release of endorphins and encephalins (related to the opiates).

Electrical techniques for producing analgesia
Transcutaneous skin stimulation.
Dorsal column stimulation.

TRANSDUCTION OF A STIMULUS AT A RECEPTOR

Application of a stimulus to a receptor cell leads to changes in

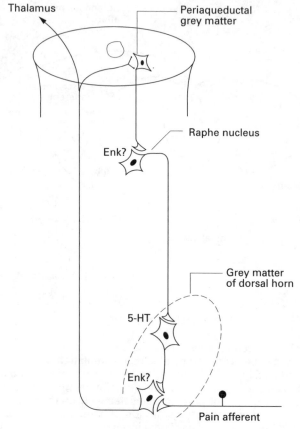

Fig. 6.4
The brain stem pain-suppressing system. Enk = encephalins; 5-HT = 5-hydroxytryptamine (putative transmitters).

conductance of the receptor membrane. This is called the receptor potential. This gives rise to a local generator potential which results in the initiation of action potentials. Local generator potentials are produced either by:
Electronic spread, or
Release of a chemical transmitter.

The receptor potential arises from opening of specific channels gated, either:
Mechanically,
Chemically, or
Thermally.

The electrical change is usually depolarisation, but may be hyperpolarisation (see vision on p. 234).

Vestibular cells produce hyperpolarisation in one direction of the hairs and depolarisation in the other.

PROPERTIES OF THE RECEPTOR POTENTIAL

The receptor potential has the following properties:
* Non-propagating.
* Has no refractory period.
* Amplitude graded according to strength of stimuli.
* Repeated stimuli summate.
* Receptor potential is more resistant to anaesthetics than is the action potential.
* The frequency of action potentials generated in the afferent nerve is approximately linearly related to the receptor potential.
  Adaptation:
* Receptor potential declines during the application of a constant stimulus.
* Slowly adapting receptors – muscle spindle, Golgi tendon organs.
* Rapidly adapting receptors – hair receptors, pacinian corpuscles.

CENTRAL CONNECTIONS OF AFFERENT NERVE FIBRES

The afferent fibres have their cell bodies (bipolar cells) in the posterior root ganglia. The central processes enter the spinal cord and each dorsal root branches into six to eight rootlets. Axons become segregated into a medial or lateral division in each rootlet.

SENSORY PATHWAYS

Consist of a chain of three neurones:
* First-order neurone synapses on cells of second-order neurone.
* Second-order neurone:
      Decussates (crosses centre line)
      Ascends in the white matter to the thalamus
      Second relay occurs in the thalamus.
* Third-order neurone runs from thalamus to somatosensory cortex.
  Information passes to higher centres by two routes within the CNS:
* The dorsal column pathway.
* The anterolateral pathway.
    The cell bodies lie in the dorsal root ganglia and their central processes pass via the dorsal root into the spinal cord.

SITE OF TERMINATION OF FIBRES IN THE CORD

The grey matter of the spinal cord has been subdivided into

laminae (Rexed laminae). The dorsal horn is divided into six roughly parallel laminae. Smaller fibres enter the cord and terminate in layers I or II. The larger fibres terminate in layers III–V.

ANTEROLATERAL PATHWAY

This pathway carries information about touch, thermal and noxious (pain) stimuli. The lateral division of the rootlets, mainly non-myelinated fibres, enter the *dorsolateral tract* (Lissauer). They terminate in laminae I, II and V at that level, but some ascend or descend up to four segments. Second-order neurones cross the cord and ascend in the anterolateral tracts. The nerves terminate:
- Ventrobasal complex of thalamus → third-order neurones (somatosensory cortex).
- Reticular formation of medulla, pons, midbrain → thalamus → hypothalamus.

DORSAL COLUMN PATHWAY

This pathway carries information concerning fine tactile discrimination, position sense and vibration sensitivity through mainly A fibres. The medial division of the rootlets enters the cord and bifurcates to enter the dorsal horn and dorsal column. Descending branches contribute to reflexes. Ascending branches pass into the dorsal columns. These ascend in the cord to gracile and cuneate nuclei and synapse with second-order neurones. These neurones decussate, i.e. cross to become the medial lemniscus, and terminate in the ventrolateral nuclei of the thalamus where they relay onto third-order neurones which pass to the somatosensory cortex and end in layer IV.

DESCENDING BRANCHES OF THE MEDIAL DIVISION

- End directly on an α-motor neurone – form part of a monosynaptic reflex arc.
- End on an interneurone → α-motor neurone (polysynaptic).

SPINAL PATHWAYS TO CEREBELLUM

- Two-neurone chain.
- Uncrossed, i.e. ipsilateral.
- Terminate in cerebellar hemispheres of the same side.

DESCENDING CONTROL OF SENSORY PATHWAYS

The cutaneous sensory pathways do not conduct impulses to the brain uncontrolled. At every synapse on the pathway there is opportunity for interaction at which:
- Irrelevant activity can be reduced.

- Discrimination can be improved.
- Contrasts can be improved.

Efferent neurones from the somatosensory cortex take part in the above processes again enhancing transmission or depressing it, or by inhibiting one channel and exciting another.

Brain-stimulated analgesia may be induced through descending pathways on stimulation of medial raphe nuclei or periaqueductal grey matter. When this is done the subject develops an indifference to noxious stimuli. This is probably due to the release of encephalins, as the effect can be blocked by a morphine antagonist (naloxone).

Figure 6.5 shows the main tracts in the spinal cord and Figure 6.6 shows the major descending tracts.

THE THALAMUS

Anatomically the thalamus is made up of a complex of nuclei, separated by laminae of white matter. In this way there are three main groups of nuclei which are further subdivided. The following abbreviations will be used:

VA = ventroanterior
VL = ventrolateral
VPL = ventral posteriolateral
CM = centromedian.

Connections of the thalamus

The thalamus is a relay station of all the ascending sensory systems except olfaction. Third-order somatosensory neurones relay signals from the lemniscal system to specific cortical areas. The thalamus also acts as a relay on the following pathways:

- Hypothalamus to limbic cortex.
- Cerebellum, basal ganglia, motor cortex.
- Retina to cortex via lateral geniculate body.
- Ear to cortex via medial geniculate body.

It is also concerned with diffusely projecting systems relating to sleep and wakefulness, mood and arousal.

Cortex and thalamus act together as a unit (Fig. 6.7)

The thalamus not only is a relay station but also receives back almost as many fibres from the cortex.

Ventral posterior nucleus
Relay nucleus for general sensation.
Fibres come from the medial lemniscus.
Spinothalamic tract.
Trigemino-thalamic tract.
Termination of the fibres is somatotrophic and modality specific.

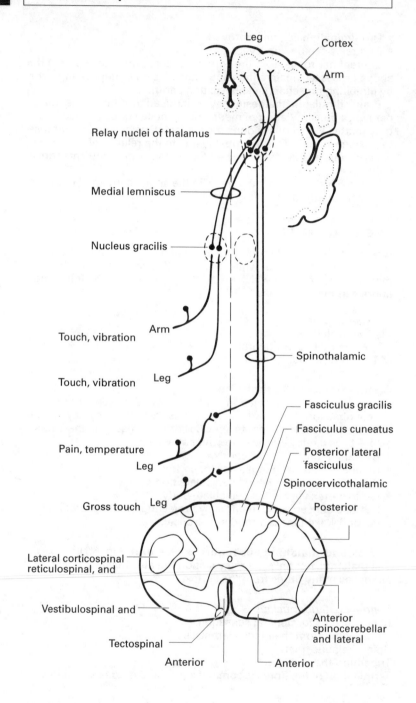

Leg

Cortex

Arm

Relay nuclei of thalamus

Medial lemniscus

Nucleus gracilis

Arm

Touch, vibration

Leg

Touch, vibration

Spinothalamic

Fasciculus gracilis

Fasciculus cuneatus

Posterior lateral fasciculus

Spinocervicothalamic

Pain, temperature

Leg

Posterior

Gross touch   Leg

Lateral corticospinal reticulospinal, and

Vestibulospinal and

Tectospinal

Anterior

Anterior

Anterior spinocerebellar and lateral

Ventrolateral nucleus
Receives fibres from:
   Cerebellum
   Basal ganglia
   Reticular formation.

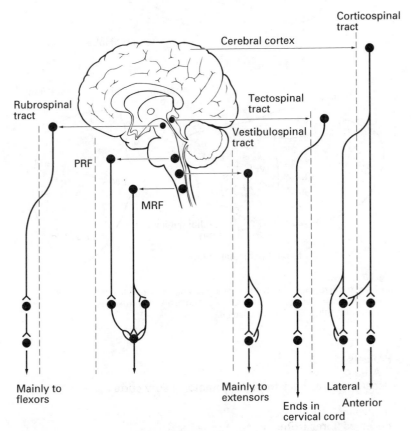

Fig. 6.6
The origins of the major descending tracts and their position in the spinal cord with reference to the midline (interrupted lines). Their positions within a cross-section of the spinal cord are shown. MRF = medullary reticular formation; PRF = pontine reticular formations; both contribute fibres to the reticulospinal tract.

Fig. 6.5
The main tracts in the spinal cord: ascending tracts on the right; descending tracts on the left.

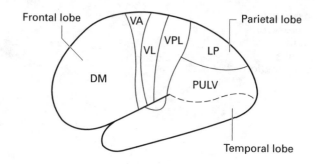

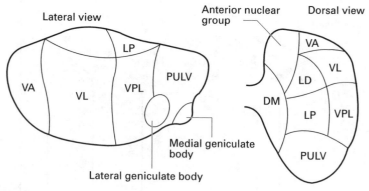

Fig. 6.7
Views of the cerebral cortex and thalamus showing corresponding regions. DM = dorsomedial; LP = lateroposterior; PULV = pulvinar; VA = ventroanterior; VL = ventrolateral; VPL = ventroposterolateral.

Projects to:
  Area 4.
Receives fibres back from the somatosensory cortex (areas 3,2,1).

Ventroanterior nucleus
Receives fibres from:
  Cerebellum
  Basal ganglia
  Reticular formation.
Projects to:
  Area 6
  Receives fibres back from areas 4 and 6.

Centromedian nucleus
Receives fibres from:

Cerebellum
Basal ganglia
Reticular formation.
Communicates widely with somatosensory and motor areas.

THE CEREBRAL HEMISPHERES

SOMATOSENSORY AREAS OF THE CEREBRAL CORTEX

Postcentral gyrus receives fibres from thalamic nuclei:
Ventroposteromedial – trigeminal
Ventroposterolateral – spinothalamic and medial lemniscal.
Electrical stimulation of the cortical areas (3,2,1) in humans gives rise to sensations of:
Touch and pressure
Occasionally warmth
Never pain.
Distribution of the sensory responses, i.e. map:
Appears as a map of the opposite side of the body
Parts of the body are topographically in the correct order
Leg is represented near to the medial edge of the hemisphere followed by trunk, neck, shoulder, arm, hand, fingers, cheeks, lips, teeth, jaw, tongue and pharynx nearest the lateral fissure.
The area devoted to each response depends upon cutaneous sensitivity and acuity, e.g. hands and lips have a larger area devoted to them than say the trunk (the sensory homunculus).
Removal of the postcentral gyrus results in an impairment of cutaneous sensation and also stereognosis. Considerable recovery may occur after a time. This is explained by:
• Ipsilateral representation.
• Sensory paths from limbs and trunk bilateral.
• Representation not confined to the postcentral gyrus.
• A secondary somatosensory area exists, at the lower end of the primary area, which receives somatosensory information from both sides.

THE ASSOCIATION AREAS

There are large areas of cortex not directly concerned with sensation and motor control which are referred to as the association areas or often the silent areas. These are:

The prefrontal cortex

Connections
Afferent from the dorsomedial nucleus of the thalamus.

Efferent to hypothalamus, striatum, subthalamus, and midbrain.
Lesions, e.g. pre-frontal lobotomy; lobotomy results in:
  Alleviation of tension and anxiety
  Changes in personality
  Useful in the treatment of intractable pain
  Little defect of intelligence except there is difficulty in carrying
  out activities that require forward planning.

The parietal cortex

Connections
Afferent fibres from:
  Somatosensory, visual, auditory and motor areas
  Thalamus pulvinar and lateral posterior nuclei
  Areas 18 and 19, the colliculi and medial and lateral geniculate
  bodies and reciprocal fibres from the parietal cortex itself via the
  pulvinar.
Efferent fibres to:
  Premotor and supplementary motor areas, the frontal eye fields,
  basal ganglia and indirect connections with the cerebellum.

Function
Most of the knowledge of the parietal lobes has come from
studying the effects of damage. Three types of disorders are
clinically recognised:
1. Agnosia – disorders of sensory analysis.
2. Apraxia – disorders of motor coordination and appropriateness.
3. Aphasia – disorders of communication.
  These disorders will include the loss of ability to:
Write.
Find one's way.
Conceive space.
Carry out commands.
Perform complex movements.
Carry out tasks such as dressing or sitting down (disturbances of
body image).

THE TEMPORAL LOBE

Some of the functions attributable to the temporal cortex now seem
to be more related to structures in the temporal cortex forming part
of the limbic system. Much of the information concerning its
function has been obtained from electrical stimulation, surgical
damage and clinical observations:

Electrical stimulation
Stimulation of the temporal cortex overlying the hippocampus
causes the patient to report fragments of past experience being
relived.

Surgical ablation
Removal of the tips of the temporal lobes produces a condition of anterograde amnesia. The memory of events before the operation is good but afterwards very poor, e.g. information can only be remembered for 10 min or less.

Clinical observations
In Korsokov's syndrome, due to chronic alcoholism – probably due to thiamine deficiency – there are degenerative changes to various sites in the limbic system and an anterograde amnesia. In most of these observations the patients were suffering from epilepsy and it may be that this condition could have modified some responses.

SPEECH

Much information has been obtained by studying speech disorders.

DYSARTHRIA

This is difficulty with articulation only and depends upon the integrity of:
- Peripheral structures, motor nuclei, nerves, muscles of lips, tongue, soft palate, larynx.
- Factors which govern airflow.
  Grammatical structure and use of language is normal.

DYSPHASIA (APHASIA)

Speech disorders are due to malfunction of speech areas in the cerebral cortex. There are three principal speech areas:

Broca's area
This is a motor speech area situated in the inferior frontal gyrus (just in front of the primary motor area devoted to the lips, tongue and pharynx). It is concerned with the coordination of motor control and the grammatical construction of speech and writing. Severely affected people understand everything but are mute or can utter only a few simple words. Mildly affected people have slow speech; they can name objects and produce small words but have difficulty with sentences – *telegram speech*.

A deaf person cannot communicate in sign language with a lesion in Broca's area.

Ninety-five per cent of right-handed people and 50% of left-handed people have the speech centre in the left cerebral hemisphere.

Wernicke's area (auditory speech area)
This area is situated behind and lateral to the primary auditory area and is larger on the left than right. It projects to and controls Broca's area.

Function
Interpretation of speech.
Forms and retains memory images of speech.

Deficit
Lacks comprehension of spoken word.
Speech is fast and incomprehensible.
Broca's area out of control.

Angular gyrus
Situated in the angular gyrus in front of the primary visual cortex.

Function
Concerned with the visual aspects of speech – interpretation of the written word.

Deficit
Lack of comprehension of the written word.
Word blindness.
Nominal aphasia, cannot find the names of objects.

MOTOR CORTEX, BASAL GANGLIA AND CEREBELLUM

CEREBRAL HEMISPHERES – MOTOR CORTEX

Motor representation
There is an orderly representation of parts of the body on the precentral gyrus, Brodmann's area 4 or MI, with the area devoted to each body part determined by the degree to which it is used for skilled manipulations. The motor area is separated from the somatosensory area (Brodmann's 3,1,2) by the central sulcus. Evidence for somatotrophic maps:
• Spread of epileptical foci – Jacksonian epilepsy.
• Electrical stimulation of exposed cortex in conscious man – Penfield.

Premotor area
The premotor area is anterior to area 4 – Brodmann's area 6. It contains no Betz cells. Stimulation activates small groups of muscle, the same as nearby area 4. It is less easily stimulated than

area 4. The effects of stimulation depend upon fibres passing to area 4, since they are abolished by:
- Cutting between area 6 and area 4.
- Removal of area 4.

Area 6 may also have a tonic inhibitory effect on area 4, as stimulation of area 6 during voluntary evoked movements leads to inhibition.

Frontal eye fields
These are situated in the front (rostral) premotor cortex and stimulation causes eye movements:
- Conjugate deviation to the contralateral side is the commonest movement.
- Sometimes turning of the head.
- Probably the primary motor area for the eyes as there are no eye movements on stimulation of area 4.
- Destruction leads to loss of voluntary movements but not reflex movement.

Connections
There are reciprocal connections with the thalamus – for connections see page 199.

Corticospinal tract
In primates the corticospinal tract (CST) originates:
30% from area 4
30% from area 6
40% from somatosensory and parietal cortex.

Only a small number (about 5%) end monosynaptically on anterior horn motor neurones, each fibre branches and exerts a strong excitatory influence on many motor neurones to innervate the fine muscle of the hand and fingers.

Note on the concept of the 'upper motor neurone'
Layer V of the cortex contains the giant Betz cells and it was thought that these gave rise to the corticospinal tract. It is now known that fibres from Betz cells form only a small proportion of the fibres in this tract, e.g. 30 000 Betz cells but 1 000 000 fibres in the CST.

An alternative name is used for CST, the *pyramidal tract*, because of the shape the fibres make on the surface of the medulla (the medullary pyramids) as they pass on their way to the spinal cord. Lesions of fibres as they pass from the cortex through the internal capsule have been traditionally designated by clinicians as involving the 'upper motor neurone', as though all the CST synapse with the anterior horn motor neurones (the lower motor neurone). This clearly cannot be the case and the signs and symptoms caused by such lesions (see p. 222) are due to damage of other fibre groups.

Afferent connections with the somatosensory area
There are rich connections with areas 3,1,2, so that the Betz cells
have a wide multimodal receptive field from:
- Skin
- Proprioceptors (joint, tendon organs and muscle spindles).

Lesions of the CST
The CST can be damaged. This results in:
- Hypotonia and weakness of muscles (flaccid).
- Coarse movement of hand and arm only.
- Loss of skilled movements, especially of the fingers – fingers
  cannot act independently.

Lesions of area 4
More severe effects than of section of CST.
Flaccid paralysis → spasticity after a time.

Problems raised by electrical stimulation of the cortex
Stimulation using extracellular electrodes affects areas not only
under the electrodes but also some distance away from them. This
is due to:
- Direct spread of electrical currents to involve more Betz cells
- Interneurones under the electrodes, which spread the influence of
  other neurones away from the actual area stimulated.
  To overcome these problems a more precise technique of
stimulating individual Betz cells by microelectrodes has been
developed, which also allows recordings to be made from the same
cells. With this technique two pieces of information accrue:
Motor
The effect produced directly by the motor neurone on single
muscles.
Facilitation or inhibition of stretch reflexes, i.e. indirectly on motor
neurones.
Sensory
By recording from the electrode while stimulating at the periphery.

Integration of somatosensory input with motor output
Motor neurones receive information from a wide range of
peripheral modalities through the nearby somatosensory areas
(3,1,2). By knowing which motor neurones are connected to a group
of muscles and what sensory information is relayed to them, then
the function of this 'afferent/efferent' unit in the CNS will be known.
Such an arrangement is necessary for:
- Precise manipulation.
- Precision grip where the force required can be judged from the
  information received from cutaneous touch and pressure
  receptors and from experience.
- Those reflexes involving the cerebral cortex e.g. the placing
  reaction.

These responses may be the reason for the motor and sensory areas of the body being in such close proximity and having rich connections between the two.

Supplementary motor area

The supplementary motor area (SMA or MII) is situated on the medial surface above the cingulate gyrus and in front of leg area of area 4. A debate exists over the function of the SMA. Evidence exists for:

- The control of movements generated internally rather than from external clues.
- Only when thinking about a motor task does an increase in blood flow in the area occur and a *readiness potential* is maximal over the SMA which is larger for an internally driven movement as against an externally cued movement.

The SMA may have a role in facilitating coordinated movement on the two sides of the body.

The motor areas in relation to speech are considered on page 205.

Volitional movements

Some cells in the posterior parietal cortex fire about 90 ms ahead of voluntary acts and not with involuntary movements. The proposed neuronal circuit is as follows:

Parietal cortex → striatum → thalamus → area 4.

THE BASAL GANGLIA

The physiological function of this complex of nuclei is still uncertain. Much of our knowledge is derived from the effects of clinical lesions in man. The afferent and efferent pathways into the basal nuclei are:

Afferent pathways

Striatum (main input structure); receives projections from:
    All regions of the cortex (excitatory)
    Intralaminar nuclei of the thalamus
    Substantia nigra.

Efferent pathways

The cortex → basal ganglia → cortex loop.
Striatum → globus pallidus → project to ventroanterior thalamic nucleus → cerebral cortex.
Striatum → substantia nigra → project to ventrolateral thalamic nucleus → cerebral cortex.

In the above neuronal circuit thalamic neurones are normally inhibited by globus pallidus and substantia nigra neurones. Thus,

an inhibition of these neurones by striatal connections will release thalamic cells from inhibition. As a result, cortical target cells are excited. Areas of cortex involved in the feedback loop are:
- SMA: MI
- Premotor areas.

Other projections include:
- Substantia nigra → superior colliculus – involved in the control of head and eye movements.
- An inhibitory projection from the globus pallidus and substantia nigra to the midbrain pedunculopontine nucleus (involved in the control of locomotion).
- Subthalamic nucleus:
     Particularly important in locomotor control
     Receives excitatory inputs from the cortex and inhibitory inputs from the globus pallidus
     Sends an excitatory projection to the pedunculopontine nucleus.

Functions of the basal ganglia (Fig. 6.8)
Our understanding of the basic functions of the basal ganglia come from clinical observation combined with pathology. Lesions of the basal ganglia lead to involuntary movements. These include:

Hyperkinetic
Choreic movements – dancing movements (corpus striatum).
Athetosis – slow writhing movements (corpus striatum).
Ballismus – flinging movements (subthalamus).

Hypokinetic Parkinsonism (nigrostriatal)
Damage to the globus pallidus and substantia nigra results in Parkinson's disease (paralysis agitans). The face is expressionless (mask-like). Tremor is the principal component present that is most obvious at rest. Tremor involves involuntary movements – *pill rolling*, alternate abduction and adduction of the thumb. Affected muscles are rigid:
- Flexor and extensor muscles alike are affected.
- Rigidity may be intermittent (cog-wheel resistance).
- Depends upon γ-efferents.
     Site of origin of tremor   Tremor disappears if patients suffer a hemiplegia or if the motor cortex is destroyed. It is also abolished by destruction of the globus pallidus or ventrolateral nucleus of the thalamus. The components of the neuronal circuit involved are:
- Basal ganglia – thalamus – motor cortex.
- Substantia nigra becomes depigmented.
- Dopamine content of corpus striatum and substantia nigra reduced.
- There is clinical improvement when L-Dopa is given by mouth.
- The disease is probably due to a degeneration of the

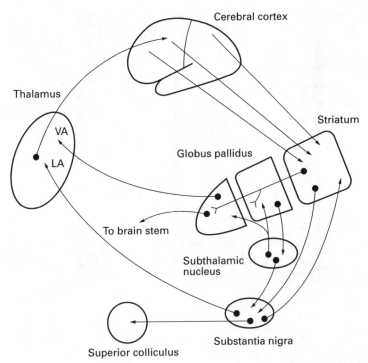

Fig. 6.8
Connections of the basal ganglia.

dopaminergic nigrostriatal pathway with consequent imbalance between excitatory (ACh) and inhibitory (dopamine) action on the striatum.

THE CEREBELLUM

Neuronal circuitry of the cerebellum
The basic structure is:

Input system
1. The climbing fibres, excitatory and precise
2. The mossy fibres, diffuse and non-specific.

Output system
The Purkinje cell.

An interneurone system
This consists of:

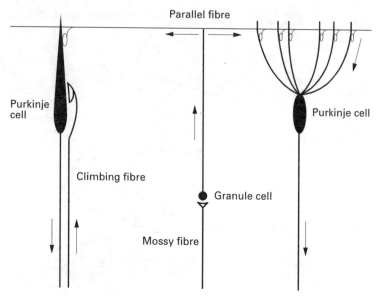

Fig. 6.9
Basal neuronal circuitry of the cerebellum.

Granule cells
Golgi cells
Basket cells
Stellate cells.

Figure 6.9 shows the two input systems, on the left the climbing fibre synapsing with the Purkinje cell, on the right the mossy fibre synapsing with the interneurone granule cell which gives rise to the parallel fibres. These in their turn run along the long axis of a folium and pass between the branches of the dendritic tree making synaptic contacts.

Figure 6.10 shows the connections of the remaining three interneurones (Golgi cell, basket cell and stellate cell).

A knowledge of the neural circuitry alone is not enough to understand the function of the cerebellum. Experimental stimulation of sensory neurones (from skin, proprioceptors, acoustic or visual receptors) produces responses in restricted regions of the cerebellar cortex where a crude somatotopical localisation can be shown. Stimulation of the sensorimotor areas of the cerebral cortex also produced cerebellar responses, confirming the neurone-anatomical evidence. Stimulation of a climbing fibre excites very few Purkinje cells but each mossy fibre stimulates a large number. Excitation of a small region of cerebellar cortex itself stimulates a small bundle of parallel fibres immediately

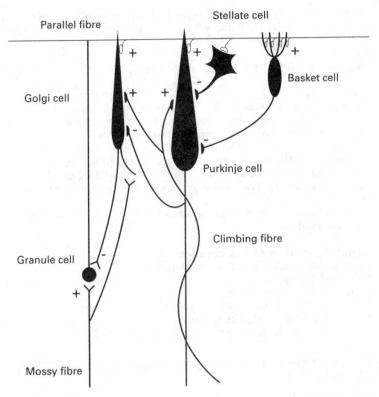

Fig. 6.10
Interneurones of the cerebellum.

under the stimulus site, and these in turn excite the underlying Purkinje cells. The Purkinje cells on either side of these are strongly inhibited by the stellate and basket cells. The boundaries of the active zone are therefore sharpened by this inhibition.

The Purkinje cells are inhibitory neurones, therefore the implication must be that an excitatory input to the cerebellar cortex is transformed into an inhibitory output to the cerebellar nuclei.

Structure of the cerebellar cortex
The cerebellar cortex is composed of three layers:
1. Molecular layer.
2. Purkinje cell layer.
3. Granular layer.

Molecular layer
The molecular layer is made up of axons running along a folium

parallel to its long axis. These axons intersect the dendritic tree of the Purkinje cells at right angles.

Purkinje cell layer
This is the layer containing the cell bodies.

Granular layer
There are several distinct cell types in the granular layer:
  Granule cells are excited by mossy fibres and send axons to the molecular layer where they form the parallel fibres.
  Golgi cells (in Purkinje layer) inhibit granule cells and dendrites and receive excitation in both molecular and granular layers.
  Basket cells (molecular layer) Dendrites excited by parallel fibres and axons form a meshwork around the initial part of the Purkinje cells.
  Stellate cells (superficial in molecular layer) Axons terminate on Purkinje dendrites.

General function of the cerebellum
The primary functions of the cerebellum are:
- Control of posture.
- Coordination of movement.
- Coordination of the cardiovascular response to exercise.
- Modifies rather than initiates function.
- Tone and movement modified on ipsilateral side.

CONNECTIONS BETWEEN THE CEREBELLUM AND OTHER PARTS OF THE BRAIN

Pontocerebellum (Fig. 6.11)
The pontocerebellum receives and sends information.

Input
Cerebral cortex → pontine nuclei → cerebellar cortex except the flocculonodular node.
Inferior olive → opposite cerebellar cortex and dentate nucleus.

Output
Purkinje cells → dentate nucleus → ventrolateral nucleus of thalamus → motor areas of cerebral cortex.
Purkinje cells → dentate nucleus → red nucleus → rubrospinal tract.

Spinocerebellum (Fig. 6.12)
The spinocerebellum receives and sends information.

Input
From the spinal cord:
  Spinocerebellar tract → cerebellar cortex.
  Dorsal columns→cuneate nucleus→cerebellar cortex.

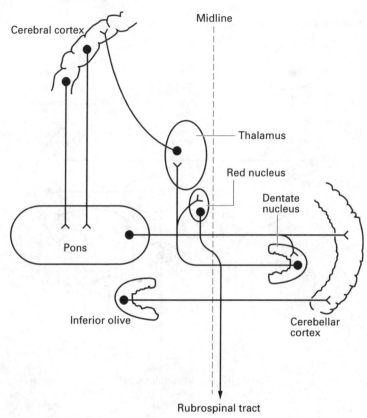

Fig. 6.11
Major connections of the pontocerebellum.

Spino-olivary tract→olive on contralateral side→cerebellar cortex
(fibres re-cross).
Vermal and paravermal cortex:
   Basal ganglia → inferior olive →cerebral cortex of contralateral
   side (vermal and paravermal cortex).

Output
Purkinje cell→intermediate and roof nuclei of cerebellum→lateral
vestibular nucleus → vestibulospinal tract.

Vestibulocerebellum (Fig. 6.13)
The vestibulocerebellum receives and sends information:

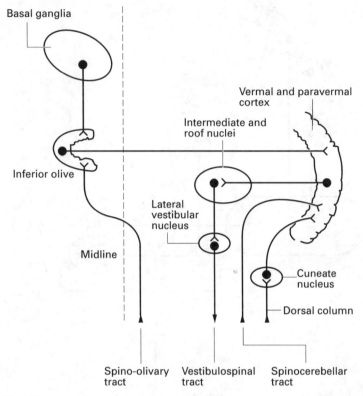

Fig. 6.12
Major connections of the spinocerebellum.

Input
Vestibule → cortex of flocculonodular lobe (direct). Not shown.
Lateral vestibular nucleus → cortex of flocculonodular lobe
(indirect).

Output
Purkinje cells of flocculonodular lobe → lateral vestibular nucleus→
vestibulospinal tract.

Afferent fibres to cerebellar cortex
They are of two types of fibre.

Climbing fibres
These fibres have their origin in the inferior olive in the brain stem
and receive input from the cerebral cortex and spinal cord, each of
which winds around its target cell and branches profusely over the

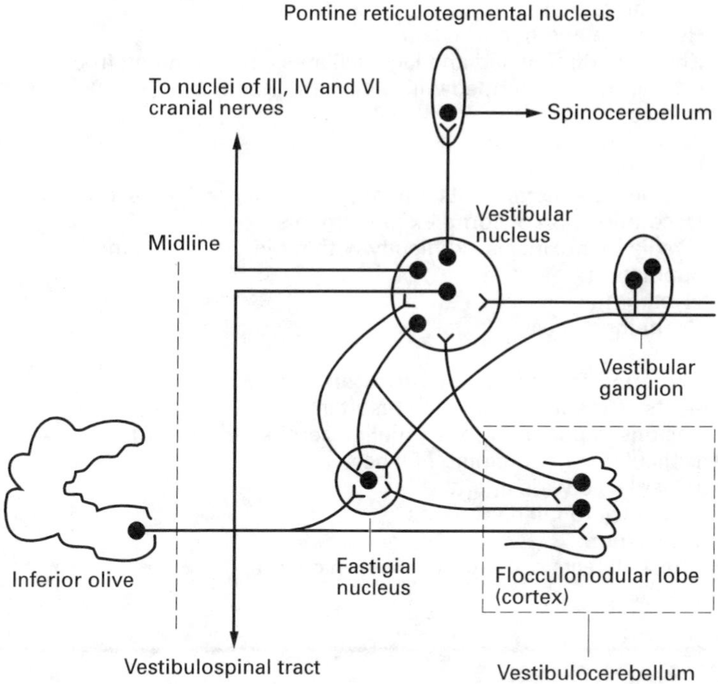

Fig. 6.13
Connections of the vestibulocerebellum.

dendrites where synaptic transmission is excitatory. They also excite deep nuclei of the cerebellum to which the Purkinje fibres also project.

Mossy fibres
The mossy fibres originate from:
- Vestibular and spinal afferents (muscle spindles) and from the auditory and visual systems.
- Nuclei in the reticular formation.
- The cortex via the pontine nuclei.

Mossy fibres enter the lower layers of the cerebral cortex and synapse in glomeruli with a number of dendrites from granular cells in the cortex. These then send ascending axons to the surface where they bifurcate → parallel fibres, forming side connections as they pass perpendicular to the planes of the Purkinje cells. In contrast to climbing fibres, mossy fibres form synapses with many Purkinje cells.

Disorders of neocerebellar function
The following may be associated with neocerebellar dysfunction:

Reflex movements:
  Hypotonia and hyporeflexia
  Knee jerk diminished and leg oscillates when hanging free
  α-Neurones undamped with γ-system underactive.
Weakness.
Voluntary movements:
  Intention tremor
  Dysmetria – movements inappropriately large, i.e. past pointing
  Decomposition of complex movements
  Dysdiadochokinesis – difficulty with rapid pronation and
  supination
  Nystagmus
  Scanning speech.

Deficiency in vestibular function

Defects in vestibular function result in:
- Lesions in posterior lobe (midline cerebellar tumour).
- Difficulty in maintaining balance.
- Feet placed wide apart.
- Tendency to fall backwards.
- No tremor.
  These defects occur particularly in children.

SPINAL REFLEXES

Reflex action

A reflex is the involuntary production of a response elicited in a
tissue or organ which is transmitted by stimulation of an afferent
fibre to the central nervous system and then to an effector organ by
an efferent nerve.

Reflex arc

The structures involved in a reflex action:
- Afferent path – receptor and afferent neurone.
- Centre – one or more synapses.
- Efferent path (final common path) – motor neurone → effector
  organ.

General characteristics of a spinal reflex action

The following components are part of a spinal reflex:
- Specificity of stimulus.
- A reflex exhibits coordination and is purposeful. For this there
  must be coordination between muscle groups, when agonists
  contract antagonists must relax.
- This requires reciprocal innervation.
  In the central nervous system interaction takes places at synapses
  which have the following properties:

- Forward conduction only.
- There is a latent response – synaptic delay.
- Facilitation occurs which is either temporal or spatial or both:
  - Subliminal fringe
  - Occlusion
  - Fractionation
  - After discharge.
- Inhibition.
- Convergence – on to the motor neurone as there is only one (final common) path.

Flexion and extension reflexes

Application of an injurious stimulus to the foot of a spinal animal results in the withdrawal of the foot on that side and the extension of the opposite limb (crossed extensor reflex).

Characteristics of flexion and crossed extension reflexes

	Flexion reflex	Crossed extension reflex
Latent period	Short (10 ms)	Long (40–100 ms)
Muscle tension	Rises suddenly	Rises slowly (recruitment)
	Declines quickly	Declines slowly (after discharge)
Stimulus for a complete tetanus	40/s to ipsilateral afferent nerve	5/s to contralateral afferent nerve
Central excitatory state	Dies away quickly	Dies away slowly

Reproduced with permission from: Bell GH, Emslie-Smith D, Paterson C 1980 Textbook of physiology, 10th edn. Edinburgh: Churchill Livingstone, p 401.)

Stretch reflexes (Fig. 6.14)

There are several types of reflex:

Phasic
The simplest reflex arc, e.g. the knee jerk.
Sudden stretch of the muscle (tap on patellar tendon).
Muscle spindle is stretched → afferent discharge.
1a afferent fibres monosynaptically excite the motor neurone to the same muscle.
Rapid contraction of the muscle.

Tonic
Characteristic of antigravity extensor muscles.
If an antigravity muscle is stretched it exhibits reflex tension. The reflex nature of the response is shown by:
  It is abolished by cutting either the anterior or posterior nerve roots

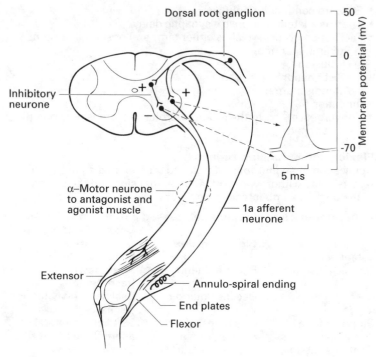

Fig. 6.14
Reflex pathways for the stretch reflex and reciprocal inhibition of an antagonist muscle. The inset shows the action potential in the excitatory alpha-motor neurone and an inhibitory postsynaptic potential in the alpha-motor neurone to the antagonist muscle.

It is inhibited by stimulating an ipsilateral sensory nerve
It is inhibited by stretching an antagonistic flexor muscle.
It is monosynaptic.

The receptor is the muscle spindle. This receptor responds to stretch and the rate of action potential discharge is increased by increasing stretch. Action potential discharge can also be activated by the γ-efferents when the muscle length is fixed. The rate of afferent discharge depends therefore on:
- Length of the muscle (rate of change of velocity).
- Length of the intrafusal fibre.
- The discharge of the γ-efferent fibres.

Muscle tone
The stretch reflex is the basis of muscle tone which is that constant muscle activity which is the background upon which the

movements and posture of the body depend. Deprivation of afferent information after posterior root section leads to:
- Loss of proprioceptive sense.
- Voluntary movements become inaccurate.

To keep tone and muscle movement in step, both α- and γ-motor neurones must be activated together. A servo-system signals to the γ-system the error between the desired length of a muscle and its actual length and sends corrective action – a short loop through the spinal cord. A force command is also necessary for smooth and accurate movements – a long loop through the cerebral cortex. Past experience and vision play an important part, together with the motor programmes of the cerebellum.

Tonic vibration reflex (TVR)
A vibratory stimulus applied to a muscle or its tendon (> 100 Hz) gives rise to sustained contraction and relaxation of the antagonist muscle.

Other postural reflexes

Positive supporting reactions
These occur in response to changes in pressure distribution and involve the conversion of a limb into a rigid pillar. There is an extensor response to the stimulus of contact with the ground. This is initiated by stretching of interosseous muscles. The vertebral column is made rigid.

In a four-legged animal pressure on the back feet only leads to rear limb extension, fore limb flexion (the animal will have a more horizontal posture if facing up a slope).

Righting reflexes
These reflexes enable a man or animal to stand the right way up. They involve:
- Labyrinthine-righting reflexes.
- Body-righting reflexes act on:
    Head
    Body.
- Neck-righting reflexes.
- Optical-righting reflexes.
- Postural reflexes involving the cerebral cortex:
    Placing reactions
    Hopping reactions.

Effects of section of the roots of the spinal nerves
Cutting or damaging nerve roots has specific actions:

Posterior roots
Cutaneous anaesthesia of the dermatomes involved.

Loss of protective pain reflexes → damage to skin and ulceration.
Loss of proprioceptive sense:
  Joint position
  Vibration
  Voluntary movements become inaccurate.
Loss of visceral sensation.
Spinal reflexes abolished.
Muscle tone reduced.

Anterior roots
Loss of voluntary and reflex movement.
Muscles are flaccid, toneless and waste.
Motor neurones of anterior horn show degenerative changes.
Autonomic effects:
  Loss of sweating
  Dilatation of skin vessels.

Comparison of 'upper' and lower motor neurone paralysis

'Upper motor neurone'	**Lower motor neurone**
1. Muscle groups are affected, never individual muscles	1. Individual muscles may be affected
2. Spastic paralysis	2. Flaccid paralysis
3. Reflexes in paralysed limbs increased; clonus	3. Reflexes absent
4. No muscle atrophy except from disuse	4. Atrophy of paralysed muscles
5. Plantar reflex – extensor	5. Plantar reflex if present – flexor
6. Electrical reaction normal	6. Response diminished or reaction to denervation
7. May have superadded 'associated movements' on attempted voluntary movement	7. No associated movement
8. Slowness of residual voluntary movement	

Transection of the spinal cord and spinal shock
After transection of the spinal cord:
- Reflexes caudal to the section are depressed for a time.
- Muscles are flaccid.
- Spinal shock lasts 2 days to 6 weeks.
  Three stages are described:
Stage 1: Spinal shock – no activity.
Stage 2: Return of reflex activity.
Stage 3: Predominant extensor activity.

Section in the thoracic region
Section of the cord in the thoracic region results in:

- Paraplegia – loss of voluntary movement of both legs.
- Anaesthesia – lower part of trunk and legs.
- Reflex activity begins first in distal parts of limbs:
    Earliest – flexor withdrawal reflex from plantar stimulation
    Flexor spasms, spread to autonomic outflow → emptying of
    bladder; bowel; sweating.
- Return of extensor reflexes – e.g. knee jerk is exaggerated; clonus
  can be demonstrated.
- Final outcome – predominately extensor activity; paraplegia and
  anaesthesia are permanent.

Hemi-section of the cord

On the side of the lesion	On opposite side
Sensory	
At the level of the lesion	
Band of cutaneous hyperaesthesia	No abnormality
If three segments are involved, band of impairment of touch, pain and temperature sensation	
Below the level of the lesion	
Touch, pain and temperature normal	Impairment or loss of pain and temperature
Postural and vibration sensation impaired or lost	Touch, postural and vibratory sensibility normal
Motor	
Below the level of the lesion	
Spastic paralysis of leg	No abnormality
Increased tendon jerks	
Clonus	
Extensor plantar response	

Decerebrate rigidity

This preparation allows some general observation to be made
concerning the tonic action of parts of the nervous system. It is
produced by a section across the brain stem between the level of
the red nucleus and the lateral vestibular nucleus.

Attitude of the animal
When the anaesthetic has worn off:
- Limbs extended.
- Tail raised.
- Head elevated.
- Pattern of stiffness in the legs depends upon the way up the
  animal is.
- It depends upon the position of the head.
- Animal rigid when placed on its feet.

- Respiration and other medullary reflexes present.
- Temperature control and voluntary movements absent.

Cause of attitude
Hyperactive stretch reflexes.
Exaggerated muscle tone is reflex in origin:
   Depends upon the integrity of the dorsal roots
   Muscles relax when tendons are cut.
The rigidity is abolished by lesions of the lateral vestibular nuclei.

Conclusions
The inhibitory influence on the lateral vestibular nucleus is lost:
- Source of the inhibition:
   Not the red nucleus
   Superior reticular formation.
- Facilitatory reticular formation has unopposed action.

Excitatory reticular formation

Extent
From the thalamus rostrally to the hindbrain caudally.
Electrical excitation causes widespread excitation of the brain.

Input
Mainly ascending from the spinal cord.
All the main sensory tracts send collaterals.

Output
Mainly upwards to the cerebral cortex; a minor component to the spinal cord.

Activation
Generalised activation of the cortex is due to volleys of impulses along ascending pathways.

Inhibitory reticular formation
Electrical excitation causes widespread inhibition of the brain.

Extent
Mainly in the caudal part of the medulla.
Projects mainly downwards to the spinal cord with little inhibitory effect to the cortex.

Input
Cerebral cortex.
Basal ganglia.
Red nucleus – the most powerful influence on muscle tone.

Output
Spinal cord.

The righting reflexes
Movements that return the body to a standing posture whenever that posture has been disturbed.

Tonic neck reflexes
The position of the neck gives rise to afferent impulses which enter the CNS by C1–C3 dorsal roots and initiates tonus changes in distant parts of the body. These are best seen in the decerebrate labyrinthectomised cat. The sequence of events is:
- Nose raised (dorsiflexion of neck) → forelimbs extended, hind limbs relaxed.
- Nose directed downwards → forelimbs relaxed, hind limbs extended.
- Head to right or ear to shoulder → right limb extended, left limb relaxed.
- Head to left → left limb extended, right limb relaxed.
- Pressure over cervical vertebrae → all four limbs relaxed.

THE LABYRINTHS

BONY LABYRINTH

Contains three communicating cavities:
- Vestibule
- Semicircular canals
- Cochlear.
  All contain perilymph with a high Na^+/K^+.

MEMBRANOUS LABYRINTH

Lies within the bony labyrinth and consists of:
- Cochlear duct
- Utricle and saccule – lie within the vestibule.
  All contain endolymph with a high K^+/Na^+.

SEMICIRCULAR DUCTS

There are three on each side. Each is at right angles to the other two. With the head erect:
- The lateral (horizontal) canal slopes downwards and backwards at 30°.
- The superior (anterior) canal is 45–55° to the sagittal plane.
- The posterior canal is also some 45–55° to the frontal plane.
- The left and right canals are in the same plane.

- The posterior canal on one side is in a plane nearly parallel to the superior canal of the other side.

  The canals connect at both ends with the utricle so the endolymph is free to circulate. The endolymph circulation is monitored by receptors situated in swellings at the end of each canal called the ampullae.

THE RECEPTORS

Projected into the ampulla is a crest, the crista ampullaris, which is covered with sensory hairs. These hairs are embedded into the *cupula*, a jelly-like flap which runs across the width of the ampulla and swings backwards and forwards in response to the fluid movements. The flap has the same specific gravity as the endolymph and therefore does not respond to gravity, but to rotation of the head about *any axis in space*, e.g. acceleration of the head to left → cupola deflected opposite to motion (due to inertia of endolymph) → stimulation from the left horizontal canal → increasing frequency of action potentials with decreasing frequency from the right side (mirror image information).

SACCULE AND UTRICLE

There are two types of sensory cell:
- Type I, flask shaped
- Type II, more cylindrical.

Structure
Confined to a sac, called the *macula*.
Cilia project from the upper surface.
At the edge of the cell is a large flexible *kinocilium*.
To its side are 60–100 *stereocilia* (thin and stiff) graded with the tallest near the kinocilium and shortest towards the other side of the cell.
Tips of stereocilia linked by filaments to its neighbour.
Cilia embedded into a jelly-like mass (the *otolith*) into which are incorporated calcite crystals (*otoconia*).

Excitation
Bending of the cilia in the direction of the kilocilium → excitation.
Bending away → inhibition.
Bending alters ionic permeability → mechanical opening of channels → current flow → membrane potential altered at end of cell → calcium entry → transmitter release (? gluamate).
Receptors fire spontaneously but bending the kinocilium one way increases the rate, bending in the other direction decreases the rate.

Innervation
The cells receive both an afferent and an efferent innervation (from

lateral vestibular nucleus), which travel in the vestibular division of the VIIIth nerve to the vestibular nuclei.

Function
The receptors respond to:
Tilting of the head → otolith moves relative to macula → bending of cilia → stimulation of afferent nerve fibres.
Linear acceleration → otolith inertia → bending of cilia → nerve stimulation.

In practice the receptors signal the effective direction of gravity which is *the vector sum of all accelerations of the head*. They also signal information about the direction of the vector as:
- The maculae of utricle and saccule lie in different planes.
- The utricle is approximately horizontal.
- The saccule is nearly vertical.

Patterns of orientation of hair cells send information about the direction of acceleration. These patterns are different in the utricle and saccule.

7. The special senses

THE EYE

THE AQUEOUS HUMOUR

The aqueous humour is the fluid within each eye. It is produced by the epithelium of ciliary processes and posterior surface of the iris. The fluid is formed by an active secretory process at a rate of 1–2 µl/min. A potential difference (PD) exists between the aqueous humour and blood. This PD is typically 6–15 mV – the aqueous humour (+) and blood (–). In relation to plasma, the composition of the aqueous humour is:

- Nearly protein free (contains about 100–200 mg/L).
- The Na^+ concentration is higher.
- The bicarbonate concentration is higher.
- The vitamin C concentration is 10–20 times higher.
- The carbonic anhydrase inhibitor acetazolamide interferes with its production.

The circulation of the aqueous humour

Fluid, formed in the posterior chamber, passes over the anterior surface of the lens through the pupil into the anterior chamber. It is removed at the iridio-corneal angle, through a trabecular meshwork. From there it moves into the sinus venosus sclerae (canal of Schlemm), from where it drains into veins within the sclera.

Function

The function of the aqueous humour is to nourish the avascular cornea and lens. It also buffers the acid produced by anaerobic metabolism of the lens. It has a mechanical function to keep the eyes rigid and to maintain the refractory surfaces. This is achieved because there is resistance to drainage which creates an intraocular pressure of 10–20 mmHg (1.3–2.6 kPa). A defect in the drainage system increases the intraocular pressure and can cause glaucoma.

THE LENS

The lens is a structure which functions to focus an external image on the retina.

Structure

The lens consists of concentric laminae of ribbon-like fibres. Its outer element, the cortex, is softer than its central region, the nucleus. The lens is enclosed in a strong membranous capsule which is thin at the posterior pole and thick anteriorly. The capsule is attached to the ciliary body by suspensory ligaments. These ligaments are kept tense by intraocular pressure, which increases the diameter of the annular ciliary body.

Metabolism

The lens functions by anaerobic glycolysis. Glucose needed for these processes is taken up from the aqueous humour.

THE IRIS

The iris forms a heavily pigmented screen which lies in front of the lens. Its inner margin forms the pupil. The diameter of the pupil is controlled by a sphincter muscle. Contraction of the sphincter has the following features:
- Reduces the pupil diameter.
- Innervated by the parasympathetic.
- Origin is the IIIrd nerve nucleus.
- Preganglionic fibres relay in the ciliary ganglion.
- Postganglionic fibres lie in the short ciliary nerves→eye.
- Major control of pupil diameter.

Radially arranged myoepithelial cells

The functions of these cells are to:
- Dilate the pupil.
- Innervated by the sympathetic pathway.
  Damage to the sympathetic pathway results in:
- Horner's syndrome.
- Miosis – small pupil.
- Ptosis (drooping eyelid) – paralysis of the smooth muscle of levator palpebrae superioris.
- Enophthalmos (eyeball retracts).

Effect of drugs on intraocular muscles

Some commonly used drugs have effects on the pupil:
- Parasympathomimetics (eserine)→constriction of pupil.
- Parasympatholytics (atropine)→dilatation of pupil.
- Sympathomimetics (amphetamine: phenylephrine)→dilatation of pupil.
- Morphine→constriction.
- Alcohol→dilatation.

PHYSIOLOGICAL OPTICS

Image formation

Parallel rays of light are focused on the retina in the normal eye relaxed for distant vision. The optical power of the eye is 60 dioptres. This is made up of the lens – 20 dioptres and the anterior corneal surface – 40 dioptres.

Accommodation

Accommodation occurs when the gaze is transferred from a distant to a near object. Images on the two retinae are blurred and disparate, i.e. they do not fall on corresponding points. This is the stimulus which causes the optical axes to converge and fall on corresponding points. This process consists of three associated events:

1. Convergence of the optical axes.
2. Constriction of the pupils.
3. Accommodation of the lens.

This is brought about by contraction of the ciliary muscles and relaxation of the suspensory ligaments. As a result of this, the anterior surface of the lens bulges forward to increase the optical power of the lens. The object is brought into focus.

Near point

The near point is defined as the nearest point to the eye at which an object may be viewed without blurring. Established long sightedness, presbyopia, is 40 cm. The near point varies with age:

- At 10 years old – 7.0 cm.
- 40 years old – 20 cm.

Errors of refraction

Defect	Parallel rays brought to a focus	Corrective lens
Myopia (short sighted)	In front of retina	Concave
Hypermetropia (long sighted)	Behind the retina	Convex
Emmetropia (normal sighted)	On the retina	—
Astigmatism	Refracting surfaces are different along different meridians	Cylindrical

The pupil

The diameter is determined by:

- Intensity of the light.
- Accommodation.

Small pupil

A pupil with a small diameter has many advantages and disadvantages for the eye:

- It reduces spherical and chromatic aberration.
- It increases the depth of focus.

- It reduces the light intensity.
- The resolving power is reduced by a reduction in light intensity.
- The optimal pupil size is determined by a balance of the above factors.

Effect of light on retina
Light falling on one eye causes constriction of that pupil; this is a direct light reflex. Constriction of the other pupil is described as a consensual light reflex.

THE RETINA

The retina is the region of the eye containing the photoreceptors and retinal nerves.

Photoreceptors

Rods
Mediate monochromatic vision.
Function in dim light – scotopic vision.
Mainly distributed peripherally in the retina.

Cones
Responsible for colour vision.
Vision in bright light – photopic vision.
Highest density in the fovae centralis.
Responsible for appreciation of fine detail.

Visual acuity
Visual acuity is the appreciation of fine detail. It is properly defined as follows:
- The least angle subtended at the eye by parallel dark and light bars when they can be just distinguished as individual bars.
- Angle subtended half a minute of arc, which is also the angle subtended by the central foveal cones which are 2 μm in diameter.

Factors affecting visual acuity
1. The position of the object in the visual field:
    a. Acuity is greatest at the fixation point – image is on central fovea
    b. Acuity is least at the periphery of the visual field.
2. The degree of illumination.
3. Contrast.
4. Optical factors:
    a. Spherical and chromatic aberration
    b. Diameter of the elements of the retinal mosaic.

Clinical measurement of visual acuity

Snellen test type   Visual acuity is the ratio of the distance of the individual from the chart to the distance at which details of the correctly read line of letters subtends an angle of 1° of arc.
Landolt's – C

Visual fields
The visual field is that portion of the external world visible out of the eye when the gaze is fixed. Clinically this is measured by perimetry. The retina has a blind spot. This is a consequence of the positioning of the optic nerve head in the retina. At this site there are no photoreceptors, hence no vision. The blind spot is usually 5–6° in diameter, 15° lateral to the fixation point. A blind spot is known as a scotoma.

PHOTOPIGMENTS AND VISION
The rods and cones in the retinal cells contain a pigment which is involved in the phototransduction of light to electrical activity (Fig. 7.1). As light falls on the pigment it results in its breakdown. This in turn leads to the generation of nervous signals. To be effective light of particular wavelengths must be absorbed by the pigment.

Scotopic vision
Scotopic vision is vision in dim light. The pigment used in this situation is rhodopsin or visual purple. A molecule of rhodopsin absorbs a quantum of light which causes a stereoisomeric change to all-trans-retinal + opsin. The rhodopsin is said to be bleached.
    The most effective wavelengths for rhodopsin are:
- Orange to violet.
- Optimal wavelength green (505 nm).
    Evidence that rhodopsin is the visual pigment:
- Absorption spectrum of the peripheral retina is identical with that of a rhodopsin solution.
- Sensitivity to dim light is greatest in the periphery of the retina.
- Rhodopsin distribution in the various regions of the retina follows closely the population density of the rods.
    With scotopic vision colour is not appreciated and visual acuity is low.

Photopic vision
Photopic vision occurs in bright light. Under these circumstances visual acuity is high and there is an appreciation of colour. Absorption spectra from outer cone segments support the concept that the cones contain three pigments:
1. Erythrolabe – in a red-sensitive cone.

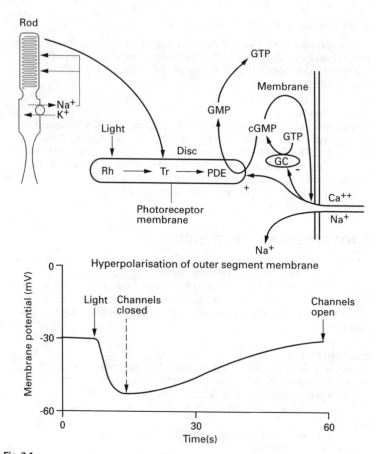

Fig. 7.1

The mechanism of phototransduction in the rods.

Information is transmitted from the photoreceptors not by action potentials but by local potentials. In the darkness the rods have open channels, maintained by cGMP, which results in a dark current which in its turn causes the tonic release of neurotransmitter on to bipolar and horizontal cells.

Light bleaches rhodopsin (Rh) to start a process which leads to the hyperpolarisation of the outer segments of the rod photoreceptor. This hyperpolarisation leads to a decease in the release of transmitter (probably glutamate). Rhodopsin is coupled to a G-protein (Tr = transducin), which activates phosphodiesterase (PDE) and converts cyclic guanine monophosphate (cGMP) to guanine monophosphate (GMP). The fall in cGMP causes the sodium channels to close and the rod membrane to hyperpolarise by diminishing the entry of Na^+ into the segment. Ca^{++} entry is also inhibited so that cytosolic Ca^{++} falls. This results in an activation of guanylate cyclase (GC) and regenerates cGMP from GTP, with the re-opening of Na^+/Ca^{++} channels. The magnitude of the activation of PDE is determined by the intensity and duration of the light stimulus.

2. Chlorolabe – in a green-sensitive cone.
3. Cyanolabe – in a blue-sensitive cone.
    Each pigment consists of a protein opsin and a chromophore retinal. There are four forms of opsin giving four photopigments.

Colour vision
The sensation of colour depends upon the spectral composition of the incident light. Any colour can be produced by mixing lights of the three primary colours, red, green and blue. There are three different groups of cones containing the three pigments. The cones are excited to different degrees according to the spectral composition of the light falling on them.

Colour blindness
Colour blindness is the inability to perceive the correct colour of an object. It is common in men (8%) but uncommon in women (0.4%). The common abnormalities are inherited as sex-linked recessive characters. There are three types of colour blindness:

Monochromats
Monochromats are rare – unable to distinguish colour at all.
Rod-monochromats:
    Have the luminosity curve of dark-adapted normals
    Lack a functional cone mechanism.
Cone-monochromats – lack a rod mechanism.

Dichromats
Protanopes – loss of erythrolabe.
Deuteranopes – loss of chlorolabe.
Protanopes and deuteranopes are red-green blind.
Tritanopes – loss of cyanolabe.

Anomalous trichromats (6% of male population)
These require three primary colours to match all colours, but require them in abnormal proportions.

Detection of colour blindness
Sufferers are often unaware of the defect and compensate with other clues. Colour blindness is detected using pseudoisochromatic (Ishihara) plates.

Dark adaptation
In a dark environment the eye increases its sensitivity. The visual threshold declines with a maximum sensitivity in about 20 min. There are two phases:

First phase
Small in magnitude.
Cones responsible.
Confined to the fovea.

Second phase
Magnitude greater.
Rods are responsible.
Sensitivity parallels the regeneration of pigment rhodopsin.
Reduced in vitamin A deficiency.

THE RETINA, VISUAL PATHWAYS AND CENTRAL CONNECTIONS

Light has to pass through layers of neurones (ganglion cells, amacrine cells, bipolar cells, horizontal cells) before reaching the photoreceptors.

The receptors
The receptors consist of two parts:
1. *An outer segment:*
   a. A high concentration of pigment
   b. A folded set of invaginations of the outer membrane containing photopigment.
2. *Inner segment:*
   a. Contains mitochondria and a nucleus
   b. Forms synaptic junctions with bipolar and horizontal cells.

The neural elements
After the reaction to light by the photoreceptors, the neural part of the retina performs an initial analysis of the information obtained. First the bipolar cells together with the horizontal cells, and then with the amacrine cells and the ganglion cells, with processes such as lateral inhibition and feedback inhibition (amacrines with bipolars) being a few of the processes being carried out. The resulting neural information is then transmitted centrally by the neurones from the ganglion cells.

Electrical response to light
Light bleaches pigment which is coupled to a G-protein (transducin). Then, through a number of intermediary steps, cGMP phosphodiesterase is activated which converts cGMP to GMP. cGMP promotes the opening of sodium channels – it requires 3 molecules per channel. In response to light the removal of cGMP effectively closes channels and reduces sodium and calcium permeability. Hence it hyperpolarises the membrane from about

–30 mV resting to –60 mV in the presence of light. Each quantum of light absorbed breaks down about 10^6 molecules of cGMP.

Photoreceptors release transmitter (glutamate) from their pedicles and when hyperpolarised (in light) they release less transmitter. The photoreceptors are functionally connected to bipolar cells.

Bipolar cells

Rod bipolar cells
Rod bipolar cells are in synaptic connection with many rods. They depolarise in response to light.

Cone bipolar cells
Cones make contact with two types of bipolar cells.
  Flat bipolars
Make conventional synaptic connections with cones.
Hyperpolarise to light.
  Invaginating bipolars
Processes from these invaginate into the pedicle.
Depolarise to light.

Horizontal cells

Horizontal cells receive information from a wide area. Their processes interdigitate with cone bipolars in the invaginations of cone pedicles. They are responsible for lateral inhibition, i.e. a bright light on the centre of a receptive field will depolarise, but when directed to the surround, will hyperpolarise.

Function of lateral inhibition
Bipolars will respond better to small stimuli in the centre of their field than to large areas which cover both centre and surround.

Detection of objects that are lighter or darker than their backgrounds is a function of the two populations of bipolar cells (hyperpolarising and depolarising).

Ganglion cells

The ganglion cells are the output cells of the retina. Unlike other retinal cells (except amacrines) they respond to light with a repetitive spike discharge. They project through fibres in the optic nerve to two major subcortical visual centres. These are the:
• Superior colliculus – concerned with eye movements and orientation to visual stimuli.
• Lateral geniculate nucleus (LGN) – projects to the cortex and is concerned with the sensation of vision.
There are many classes of ganglion cells, in the cat 23 have been described. They have receptive fields (like bipolar cells) with a

central region and an antagonistic surround. Some have an 'off' – 'on' response to changes in illumination, others respond only transiently when the level of illumination changes, still others show a sustained response to steady illumination over a wide field and some responses are colour coded, others not.

Amacrine cells
There are a large number of distinct types with many different synaptic transmitters.

Amacrines form an intermediary between rod bipolars and ganglion cells, i.e. all signals in the dark-adapted state must be transmitted to ganglion cells via amacrines. Much of their function remains unknown.

Figure 7.2 shows the distribution of rods and cones in the retina.

THE VISUAL CORTEX

The primary visual cortex is area 17 (Brodman), known also as the striate cortex. It represents the whole contralateral visual field. This region of the brain is surrounded by area 18 – pre-striate – and that by area 19 – medial temporal. In a new classification area 17 = V1, area 18 = V2–V4 and area 19 = V5. There are direct interconnections between these areas and indirect ones through the pulvinar of the thalamus.

Ablation of the visual cortex results in blindness. However, the reaction of the pupils to light is unaffected.

Representation
The right cortex receives information from the left visual field of each eye and vice versa. Along the postcalcarine sulcus, the centre of the visual field is at the tip of the occipital lobe, with the lower half of the visual field (upper half of retina) on the upper part of the primary visual cortex; the upper half of the visual field on the lower part of the visual cortex. The fovea has a greater representation than the peripheral retina.

Information comes to the cerebral cortex through fibres from the lateral geniculate nucleus (LGN). The LGN has six layers with two major subdivisions:
1. Two ventral *magnocellular* layers (1 and 2) – project mainly to superior colliculus and other subcortical areas, but some fibres do go to the cortex area 4Cα (VI). This system processes information concerned with movement, texture, stereoscopic vision but probably not colour.
2. Four dorsal *parvocellular* layers (3–6) – project to the cerebral cortex (4Cβ). These layers receive inputs from either the ipsilateral eye (layers 2, 3 and 5) or from the contralateral eye (layers 1 and 2).

There are two subdivisions:

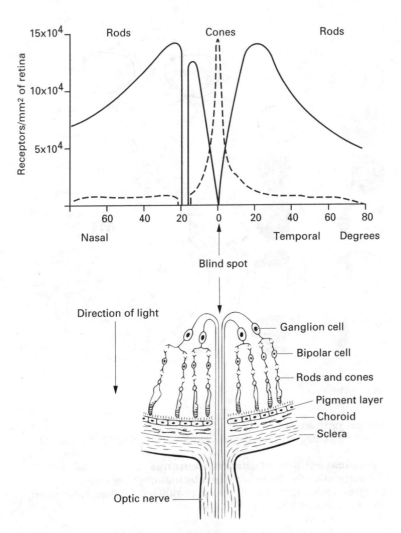

Fig. 7.2
The distribution of rods and cones in the retina.

   a. Projections to blob* regions of visual cortex – concerned with colour
   b. Projections to interblob regions of visual cortex – high spatial resolution and orientation selectivity.
  (*A blob region contains a high concentration of cytochrome oxidase.)

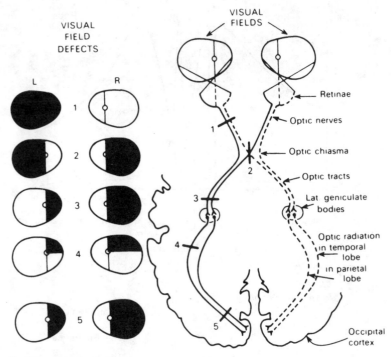

Fig. 7.3
The visual pathway

Visual pathways
Effect of damage on the visual field is summarised in Figure 7.3.

Electrical activity of cortical neurones
A neuronal hierarchy exists whereby neurones require progressively more specific visual stimuli before they are excited. The cells are classified as follows:
1. *Simple cells*
   a. Orientation detectors – respond best to a line.
2. *Complex cells*
   a. Respond best to stimuli (bar or edge) of specific orientation.
   b. Not sensitive to position in the visual field.
   c. Moving stimuli are more effective.
3. *Hypercomplex or end-stopped complex cells*
   a. Respond only if lines are within certain lengths and orientations, i.e. stimulated by contours only if they end within their receptive field.

The visual cortex appears to be functionally grouped in columns perpendicular to the surface of the cortex. In each column, the cells share the same preferred orientation. As recordings are made across the cortical surface the orientations change systematically, until after a very short distance (mm) the original orientation is again obtained. In this way every possible orientation is represented.

Electrical stimulation of area 17

In a person whose retinal or visual pathways are damaged, electrical stimulation of area 17 produces a subjective sensation of light – a phosphene. The brightness of the sensation depends upon strength of stimulation and it appears to be brighter at the beginning of stimulation. It often fades in 10–15 s, if stimuli are prolonged. The apparent position depends upon the part of the cortex stimulated.

VISUAL REFLEXES

Light reflex
Bright light directed on to the retina gives rise to:
- Pupillary constriction.
- On same side – direct reflex.
- On opposite side – consensual reflex.
  Reflex pathway involves:
- Photoreceptor activation.
- Activity in the optic nerve.
- Information relay in pretectal region.
- Edinger–Westphal nucleus of each side.
- Preganglionic fibres.
- Ciliary ganglion.
- Ciliary nerve (postganglionic parasympathetic fibres).
- Constrictor pupillae.

Light reflexes are intact after visual cortex ablation but are abolished after lesions in the pretectal region (Argyll-Robertson pupil). Constriction of accommodation is unaffected by pretectal damage.

Movements of the eye
The eye is moved by extraocular muscles:

Movement	Muscle	Innervation
Rotation of eye around vertical axis	Medial rectus	III nerve
	Lateral rectus	VI nerve
	Superior oblique	IV nerve
Rotation of eye around an oblique axis	Inferior oblique	
	Superior rectus	III nerve
	Inferior rectus	

Muscle	Direction of eye movement
Lateral rectus	Abduction
Medial rectus	Adduction
Superior rectus	Elevation, plus adduction and rotation of upper part of pupil towards the nose
Inferior rectus	Depression, plus abduction and rotation of upper part of pupil away from the nose
Superior oblique	Depression, plus abduction and movement of upper margin of pupil towards the nose
Inferior oblique	Elevation, plus abduction and rotation of upper margin of the pupil away from the nose

Control is through the midbrain nuclei (III, IV and VI).
- Eyes move together:
    Conjugate movements
    Convergence.
- Objective to produce retinal images on corresponding points of the retina. Failure leads to:
    Diplopia (double vision)
    Squint→double vision→suppression of one image.

Fixation movements
The eyes are never at rest and eye movements can be categorised:
- Saccadic movements (e.g. in reading).
- 'Optico-kinetic nystagmus' – smooth pursuit movements.
- Vestibular movements – rotation of the head→nystagmus.

There is a micro-tremor during fixation which is necessary for normal vision. When it is eliminated a stable image on the retina fades. The fixation reflex is present at birth but is weak. In early life eye movements are independent and uncoordinated. Conjugate fixation is established in 5–6 weeks. Voluntary fixation is in place by 3 months.

THE EAR AND HEARING

The ear is made up of three components:
1. The pinna and external auditory meatus.
2. The middle ear.
3. The inner ear.

EXTERNAL AUDITORY MEATUS

The external auditory meatus is about 30 mm long and conducts variations in air pressure from the pinna (auricle) to the tympanum (eardrum).

The tympanic membrane and auditory ossicles
The bones of the inner ear are the malleus, incus and stapes (auditory ossicles). They function as a matching device to transfer sound energy from a region of low impedance (air) to one of high impedance (fluid). As sound waves strike the tympanic membrane, movement is initiated which is communicated to the handle of the malleus. This causes the long process of the incus and the stapes to move in the same direction as the handle of the malleus. A relatively large movement of the drum is converted by lever action of the ossicles, into a smaller more forceful movement of the stapes. This action is reinforced by the stapes rocking on a fulcrum at the oval window. In this way vibrations are transmitted to the perilymph of the scala vestibuli.

MIDDLE EAR

The middle ear is transgressed by the auditory ossicles. It is linked to the pharynx by the pharyngo-tympanic tube. This is closed at the pharyngeal end and opened during swallowing. Air pressure is equalised across the tympanic membrane. An increased pressure during diving leads to a loss of some hearing acuity and maybe pain. A decrease in pressure at high altitudes is similar to the case described above. Hearing is restored or pain relieved by swallowing. This equalises pressures. Hearing acuity is lost or diminished if tube is blocked by swelling or mucus during a common cold.

INNER EAR

The inner ear contains the cochlea which is contained in bony tube two and three-quarter turns around the bony modulus. The cochlea is divided by Reissner's membrane and at the basilar membrane (with the organ of Corti) into three compartments:
1. The scala vestibuli above.
2. The scala tympani below.
3. The scala media between.
   The scala vestibuli is continuous with the scala tympani at the helicotrema and contains perilymph with the composition of extracellular fluid (ECF). The cochlear duct is a closed tube and contains the endolymph which has a composition similar to the intracellular fluid (ICF). Scala vestibuli begins at the oval window into which fits the foot of the stapes. The scala tympani ends in the round window. The basilar membrane stretches from the bony spiral lamina to the spiral ligament and supports the organ of Corti.

The organ of Corti

In the organ of Corti the sensory cells are the hair cells. Hair cells are held rigid by the reticular lamina and project into the gelatinous tectorial membrane. There are three rows of outer hair cells separated by a tunnel from one row of inner hair cells. Each outer hair cell carries about 120 hairs and the inner cells about 50–60. The VIIIth or auditory nerve has its cell bodies in the spiral ganglion located in the bony modiolus. At the base of each hair cell are the afferent terminals of the auditory nerve. Those to the outer hair cells cross the tunnel of Corti. The central processes pass to the cochlear nucleus.

Mechanical aspects of cochlear function

The basilar membrane

Sound waves activate the stapes and the vibrations which are set up travel as waves in the perilymph of the scala vestibuli. The behaviour of the travelling wave is determined by the properties of the basilar membrane:

- Basal end is stiff – apical end is less stiff.
- Basal end is narrow – apical end is broader.

Vibrations are set up in the perilymph which imparts to the basal membrane a wave-like like movement which travels up the cochlea increasing in amplitude slowly. After the maximum displacement, with a distance from the base that depends upon the frequency of the sound, the displacement falls off sharply. At high frequency sound crosses the basilar membrane near its base (at point of least resistance) to the round window. Low frequencies, on the other hand, are able to overcome the membrane stiffness and the inertia of the perilymph and move the membrane more easily at the apical end.

Behaviour of the cochlear partition (basilar membrane plus organ of Corti)

As well as being displaced as described above, the basilar membrane also bends about an axis near to the modiolus. This sets up a shearing motion between the tectorial membrane and the reticular lamina, bending the cilia to one side or another.

The sensory elements (Fig. 7.4)

The inner (IHC) and outer (OHC) hair cells have different synaptic connections, with about 10% of the afferent nerve fibres coming from the OHC and 90% from the IHC. Both IHC and OHC probably release glutamate. The hair cells also receive an efferent innervation, direct on to the OHC but presynaptically to IHC, which is probably cholinergic and comes from the superior olivary nucleus.

Electrical activity of the cochlea

The electrical events in the cochlea occur in the following sequence:

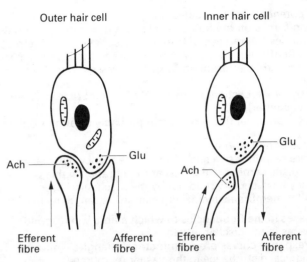

Outer hair cell   Inner hair cell

Ach   Glu   Ach   Glu

Efferent   Afferent   Efferent   Afferent
fibre   fibre   fibre   fibre

Fig. 7.4
Innervation of inner and outer hair cells. As well as feeding into an afferent neurone
(glutamate (glu) may be the transmitter), the hair cells also receive an efferent
innervation from the superior olive. They terminate directly on the outer hair cells, but
presynaptically on the inner hair cells where the transmitter is probably acetycholine
(ACh).

1. The hair cells have resting potentials which are −40 mV negative
   to the scala tympani.
2. A steady potential (the endocochlear potential) of $\cong$ 80–90 mV
   exists between endolymph and perilymph (+ve). This is
   maintained by a Na^+/K^+ pump located in the stria vascularis. The
   potential is:
   a. Oxygen sensitive
   b. Abolished by cyanide
   c. Maintained by metabolic activity of the stria vascularis.
3. The consequence of these two potentials is that they sum across
   the cochlear duct and the hair cell membrane, so that any
   conductance change across this membrane will give rise to a
   bigger generator current.
4. Cochlear microphonic potential is a potential fluctuation
   recorded between the cochlear duct and the scala tympani (in
   fact anywhere in the vicinity of the cochlea).
   The electrical equivalent to the acoustic stimuli resembles the
waveform of the stimulus. The potential has virtually no latent
period and increases in amplitude with the energy of the sound
energy. It is thought to be the summed generator potentials of large
numbers of hair cells.

Action potentials in the cochlear nerve
Shearing forces on the hair cells activate generator potentials. These induce the release of transmitter substance – glutamate. As a result afferent nerve endings are stimulated. These in turn are influenced by efferent endings from the superior olivary nucleus.

- OHC by a direct action on the basal cell membrane (ACh probably the transmitter).
- ICH by a presynaptic action (again probably through ACh).

Response of nerve fibres
Each fibre responds to a characteristic frequency (i.e. maximum point of displacement of the basilar membrane). In this way, a frequency analysis is carried out by the mechanical properties of the basilar membrane. Increasing stimulus strength results in:

- Increases in the frequencies to which a fibre will respond.
- Increases in the discharge rate.
- The response curves are asymmetrical triangles – mirror mechanical disturbance to the basilar membrane.

Nerve fibres from the basal end of the basilar membrane have low characteristic frequencies. Nerve fibres from the apical end of the basilar membrane have high characteristic frequencies. The selectivity is sharpened by lateral inhibition, e.g. the tuning curve to a single auditory unit is recorded using a pure tone and then if a second tone is added, the regions in the near neighbourhood of the first response are narrowed.

The auditory pathway is illustrated in Figure 7.5.

Central connections of the cochlear nerve
Fibres leave the cochlea arranged in order of frequency sensitivity. They terminate in the cochlear nuclear complex. Second-order neurones arise from:

- Dorsal region→contralateral inferior colliculus.
- Ventral region→superior olivary nucleus→inferior colliculus.

The two paths recombine at the inferior colliculus. The superior olive is the first area where information from one ear meets the other. From the inferior colliculus fibres go to the medial geniculate nucleus (both crossed and uncrossed fibres are directed to the primary auditory cortex, superior temporal gyrus). The auditory cortex corresponds to Brodmann's areas 41 and 42. The representation in the primary auditory cortex is tonotopically arranged. Secondary auditory cortex representation is of more complex sounds.

Temporal lobes in humans
Clinical deafness does not result from lesions of the temporal lobe unless these are bilateral and very extensive. Unilateral lesions can disrupt the patterned sequence of tonal stimuli.

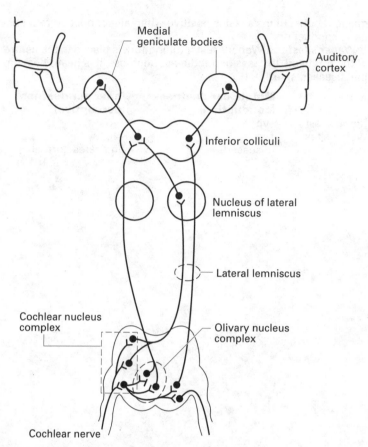

Fig. 7.5
The auditory pathway.

Methods of measuring acuity of hearing

Quantitative measurements
Using the pure tone audiometer.
- Threshold of audibility:
    Lowest (acuity greatest) at 1000–4000 Hz.
- Range of audibility:
    30–20 000 Hz in the young
    Upper limit in the old of 5000 Hz.
Zero mark on the audiogram is the average threshold in the young.

Intelligibility tests
    Rinne's test Tuning fork placed next to external meatus, then on

mastoid bone. Rinne's test is positive when air conduction exceeds bone conduction.

Weber's test In Weber's test a tuning fork is placed in the centre of the forehead. In cases of middle ear deafness, it is heard best on the diseased side.

	Middle ear deafness (conduction)	**Inner ear deafness (perceptive)**
Rinne's test	−ve	+ve
Weber's test	+ve (diseased)	−ve (indicates normal)

8. The endocrine system and reproduction

HORMONES

The word 'hormone' was first recorded in 1905 and it arose from the discovery of secretin by Bayliss and Starling, who had previously written 'The peculiarity of these substances is that they are produced in one organ, and carried by the blood current to another organ, on which their effect is manifest'. Recently it has become clear that this definition applies to the 'classical hormones', but now needs some modification. Hormones can act so that carriage in the blood is not necessary. They can:

- Act upon contiguous cells – a *paracrine* function.
- Modify the activity of cells which produce them – an *autocrine* function.

THE CLASSICAL HORMONES

Arise from glands in different areas of the body.
Each gland produces hormones with a different function.
A single hormone can exert its effects on different tissues.
A single function can be regulated by more than one hormone.
Release their products directly into the blood stream, i.e. the glands are ductless.

CHEMICAL NATURE OF HORMONES

Hormones fall into three general groups:

1. Derived from a single amino acid:
   a. Noradrenaline, adrenaline and dopamine
   b. Thyroxine derived from two iodinated tyrosine molecules.

2. Derived from peptides or proteins:
   a. Those first discovered in the gastrointestinal tract and varying in size from a few amino acids to those with 200 or more, e.g. from TRH to HG and FSH.

3. Derived from cholesterol, the steroids of which there are two types:
   a. Intact steroidal nucleus – adrenal steroids
   b. Broken nucleus – vitamin D and metabolites.

MODE OF ACTION OF HORMONES

The initial step in hormone action is its binding to a receptor. For protein, peptide and amine hormones, the receptors are on the cell membrane. For steroid and thyroid hormones, the receptors are in the cytosol.

For membrane receptors the next stage is the formation of second messengers. The possible proposed candidates are:

Ca^{2+}

cAMP←————————Adrenaline, vasopressin, ACTH, LH, TSH,
                         glucagon parathyroid hormone, secretin, etc.

cGMP

Prostaglandins

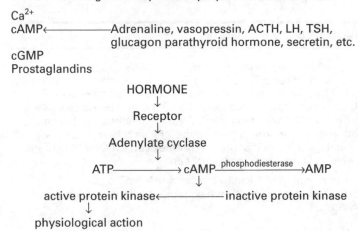

Sterols

Steroid and sterol hormones (lipophilic ligand) enter cells and bind to receptors in cytosol. The hormone–receptor complex enters the nucleus and binds to DNA where it activates or suppresses the transcription of specific genes. This may involve splicing of exon sequences, leading to the production of specific messenger RNA. As a result there is an increase in the production of specific cellular proteins.

Other hormones acting on protein synthesis

These include thyroxine and tri-iodothyronine. These hormones react with receptors already bound to DNA and trigger the synthesis of specific RNA and proteins.

THE ADRENAL GLAND

The adrenal gland is situated above the kidney and functions as two units, the cortex and medulla.

THE ADRENAL CORTEX

Structure of the adrenal cortex
Histologically there are three zones:
Outer – *zona glomerulosa* – secretes aldosterone.
Middle – *zona fasiculata* – secretes mainly glucocorticoids.
Inner – *zona reticularis* – secretes mainly androgens.

Synthesis of adrenal steroids
Adrenal steroids are derived from cholesterol, which is either
synthesised within the adrenal cells or is taken up from low density
lipoproteins in the blood.

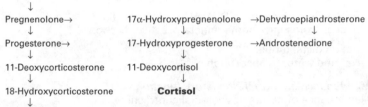

Cholesterol
↓
Pregnenolone→   17α-Hydroxypregnenolone   →Dehydroepiandrosterone
↓   ↓   ↓
Progesterone→   17-Hydroxyprogesterone   →Androstenedione
↓   ↓
11-Deoxycorticosterone   11-Deoxycortisol
↓   ↓
18-Hydroxycorticosterone   **Cortisol**
↓
Aldosterone

Storage
Hardly any storage.
Rate of release approximates to rate of synthesis.
About 30–90 mmol (10–30 mg)/day is secreted in the adult.
Corticosterone about 3 mg/day and aldosterone about 0.2 mg/day.

Transport in blood
Mostly bound to plasma proteins – inactive.
75–80% bound to α_2-globulin – transcortin.
15–20% bound less tightly to albumin.
5% free – circulating unbound.

Plasma levels
A circadian rhythm exists in plasma levels of these hormones.
- Highest in the morning before waking; lowest in the early hours
  of sleeping.
- Levels are secondary to those of ACTH and CRF.
- Secretion episodic – 7–15 brief episodes a day.
- ACTH secretion suppressed by high blood levels of cortisol.
- Buffer mechanism in the release of cortisol:
    Mopped up by the proteins at high plasma levels of the free
    hormone
    Released at low plasma levels.
- Corticosterone is qualitatively similar but secreted at one-tenth of
  the amount of cortisol.

Function of glucocorticoids
The glucocorticoids have the following major functions:

Effects on metabolism
Play a role in carbohydrate metabolism.
Promote synthesis of glycogen in the liver.
During fasting they stimulate gluconeogenesis→glucose for brain:

> Accelerates protein catabolism, e.g. in skeletal muscle
> Rise in blood glucose concentration (counter-regulatory to insulin)
> Inhibit glucose uptake in muscle and adipose tissue.

Regulate the synthesis of specific hormones.
Have a permissive action on the effects of some hormones:

> Adrenaline, glucagon, GH, ADH, angiotensin II.

Glucocorticoids have some mineralocorticoid action:

> Act on distal tubule
> Na^+ and water re-absorption.

Maintenance of normal circulatory function
Maintenance of normal myocardial contractility.
Play a permissive role in potentiation of the vasoconstrictor effects of catecholamines.
Decrease the permeability of vascular endothelium.

Adaptation to stress
Stress releases cortisol.
In adrenocortical deficiency, stress in the form of trauma, heat, cold, etc, less well tolerated.
Large doses suppress the immune response.
Also have anti-inflammatory and anti-allergic properties.

Control of secretion
Plasma levels of cortisol secretion controlled by:

> Plasma levels of ACTH from the pituitary
> ACTH secretion regulated by hypothalamic secretion of CRH into pituitary portal system
> Cortisol exerts a negative feedback on hypothalamus and anterior pituitary.

Mode of action of ACTH
Binds to specific receptors on the cell membrane.
Activates adenylate cyclase→cAMP→activation of protein kinases.
Ribosomal protein synthesis.
Synthesis of cortisol from cholesterol.

Mineralocorticoids

Aldosterone
Salt-retaining steroid.

Produced by cells of the zona glomerulosa.
Largely independent of ACTH.
Enters target cell and binds to a specific protein.
Migrates to nucleus where it promotes the synthesis of a specific protein which is associated with sodium transport.
Acts on distal convoluted tubule and collecting ducts→increases absorption of Na^+ and the secretion of K^+ and H^+, so that:

Plasma volume increases
Blood pressure rises.
See page 297.

Control of aldosterone secretion
Secreted by the zona glomerulosa.
Principal stimulus angiotensin II.
Minor stimuli – decreased sodium concentration acting through a:
Reduction in plasma osmolality
Increased plasma $[K^+]$
Reduction in ACTH.

Aldosterone secretion rate
Salt-restricted diet – 0.9–3.0 µmol/24 h.
High salt diet – 0–0.4 µmol/24 h.
Increased in oedematous states:

Cardiac, renal and hepatic disease→secondary aldosteronism
Tumours of adrenal cortex→primary aldosteronism.

Disorders of the adrenal cortex
There are several diseases associated with dysfunction of the adrenal cortex. These include:

Addison's disease
Addison's disease results as a consequence of the destruction of the cortex of the adrenal gland. This can be caused by:

- Autoimmune disease
- Tuberculous infection.
  Clinical features
Deficient secretion of all adrenocortical hormones.
Weakness and lassitude.
Loss of weight.
Hypotension.
Sensitivity to insulin→hypoglycaemia.
Excessive loss of Na^+ and Cl^- in the urine.
Hyperpigmentation due to excessive ACTH secretion→melanocyte activity.
Elevated ACTH is brought about by removal of negative feedback by cortisol.

Cushing's syndrome
Cushing's syndrome is associated with a collection of signs and symptoms, due to prolonged elevation of plasma glucocorticoids.
   Clinical features
Obesity.
Glycosuria.
Hirsuitism.
Osteoporosis.
Hypertension.

THE ADRENAL MEDULLA

The medulla is derived from the neural crest and has a common origin with ganglionic cells of the sympathetic nervous system.

Structure

The adrenal medulla consists of masses of polyhedral cells separated by large blood sinuses. Cells stain brown with ferric and chromic acid stains, hence the term 'chromaffin' cells. There is evidence that there may be two distinct cell types releasing noradrenaline and adrenaline.

Innervation

The adrenal medulla is innervated by myelinated preganglionic fibres from the greater splanchnic nerve. Adrenal cells (chromaffin cells) correspond to postganglionic cells. These cells contain two catecholamines:

- Noradrenaline (NAD) – 20%.
- Adrenaline (AD) – 80%.
  They also contain:

- Enkephalins, dynorphins, neurotensin, somatostatin and substance P, role not yet clear.
- Met- and leu-enkephalins are co-secreted with adrenaline.
  These hormones are stored as membrane-bound granules and bound to ATP and protein.

Synthesis

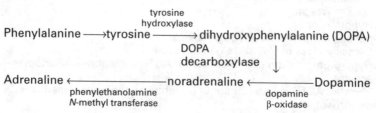

Mechanism of secretion

The release of adrenal medullary hormones occurs in the following sequence:

- Impulses in preganglionic sympathetic fibres release acetylcholine (ACh).
- Chromaffin cells depolarised.
- Increase in membrane permeability.
- Ca^{2+} enters cell (stimulus-secretion coupling – second messenger).
- Exocytosis of granules triggered.
- Catecholamines, ATP, and protein released into blood.

Action of catecholamines

See Chapter one.

Plasma levels

The plasma levels of catecholamines can be measured by the following methods:

- Chemical
- Biological assay
- Radioimmunoassay.
  Typical levels are:
- NAD 1.5 nmol/L
- AD 0.3 nmol/L.
  Plasma levels are lower in sleep.

Control of secretion

Secretion initiated through the sympathetic nerves is controlled by the hypothalamus. There is some evidence of selective secretion of NAD and AD.

Effects which increase the AD/NAD ratio
Emotional states which are unfamiliar
Pain
Hypoglycaemia
Haemorrhage
At the start of exercise.

Effects which reduce the AD/NAD ratio
Asphyxia and severe hypoxia.
Emotional states to which the subject is familiar.
Reflex activation of the sympathetics via the baroreceptors.

Electrical stimulation of hypothalamus
Areas can be found which stimulate AD secretion
Other areas can be found which stimulate NAD secretion.

Function

Adrenaline
Causes vasoconstriction in the skin
Produces vasodilatation in muscle
Stimulates metabolism.

Noradrenaline
Causes general vasoconstriction (except coronary arteries) and an increase in peripheral resistance.
Responsible for maintenance of blood pressure.
  This is a simplified classification of action, the truer situation is more complex. Before NAD and AD can act they bind to cell membrane α- and β-adrenoreceptors and their various subclasses which differ in affinity from tissue to tissue.

Disorders of the adrenal medulla

Hypofunction – no clear-cut clinical syndrome; medulla not essential to life except in severe stress

Hyperfunction – phaeochromocytoma (tumour of the chromaffin cells)
Tumours of the adrenal medulla:
  Excessive noradrenaline secretion
  Periodic elevation of the blood pressure
  Periodic bouts of sweating.

Degradation of catecholamines
Figure 8.1 illustrates the degradation of catecholamines.

THE PITUITARY HORMONES

BLOOD SUPPLY TO THE PITUITARY GLAND

The blood supply to the pituitary gland is derived from the internal carotid artery. The posterior lobe is supplied from inferior hypophyseal arteries. The median eminence and pituitary stalk are supplied from the superior hypophyseal arteries. In the anterior lobe there is a portal system from capillaries on the median eminence and pituitary stalk. The pituitary portal system:

• Provides nutrition for the gland.
• Hormones liberated from hypothalamic neurone terminals enter

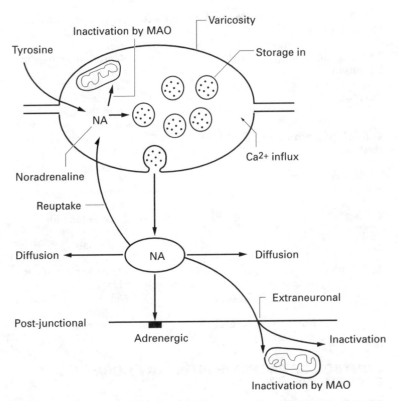

Fig. 8.1
Degradation of catecholamines. COMT = catechol-*O*-methyltransferase; MAO = monoamine oxidase; NA = noradrenaline.
The action of noradrenaline is terminated in three ways:
1. Active re-upake of the transmitter into the presynaptic ending which can amount to about 70% of the amine released. There it is repackaged into vesicles and some of the free noradrenaline is inactivated by mitochondrial MAO.
2. Extraneuronal uptake with subsequent breakdown by mitochondrial MAO or by COMT.
3. By diffusion away from the synaptic cleft. It is this which causes the so-called spillover into blood which is used as an indicator of the magnitude of sympathetic activity.

the capillary plexus of the median eminence and are carried in the portal vessels to the anterior pituitary.

PRIMARY PITUITARY HORMONES

The following table describes the primary pituitary hormones.

Pituitary hormone	Control of release	Target and action
Growth hormone	GHRH, GHIH	Growth and metabolism
Thyroid-stimulating hormone	TRH	Secretion of thyroid hormone
Adrenocorticotrophic hormone	CRH	Secretion of corticosteroids from the adrenal cortex
Luteinizing hormone (LH)	GnRH	Stimulates ovulation and testosterone secretion
Follicle-stimulating hormone	GnRH	Stimulates follicular growth or spermatogenesis
Prolactin	PRH, PIH	Stimulates milk secretion and maternal behaviour
Melanocyte-stimulating hormone	MSH, MIH	Human role not yet clear
Oxytocin	Neural	Milk ejection, uterine contraction
Antidiuretic hormone (ADH) or vasopressin	Neural	Water retention

CONTROL OF ANTERIOR PITUITARY HORMONE SECRETION

Anterior pituitary hormone secretion is regulated by neurohormones originating from neurones of the hypothalamus. They are released at the level of the median eminence into fenestrated capillaries. From there they are carried in blood through large portal vessels on the pituitary stalk to a second set of capillaries in the anterior pituitary (a portal system). In the anterior pituitary they either stimulate or inhibit the release of a corresponding hormone.

Releasing hormones

Growth hormone-releasing hormone (GHRH)
This hormone is produced in the ventromedial hypothalamus (an area associated with the control of eating).

   Stimulated by:
Hypoglycaemia
High blood amino acid levels
Onset of deep sleep
α-Adrenergic pathways
Serotonergic (5-HT) pathways.

Inhibited by:
GHIH (somatostatin) inhibits synthesis and secretion of growth hormone
Dopamine inhibits growth hormone release. There is an interaction between GHRH and somatostatin, as GHRH secretion increases as that of somatostatin decreases and vice versa. Release is pulsatile, particularly during the early hours of sleep. The frequency of release is greater in children and adolescents. It is usually absent after 50 years of age.

Thyrotropic-releasing hormone (TRH)

Thyrotropic-releasing hormone is a tripeptide which stimulates the release of TSH (thyroid-stimulating hormone). TSH release is mediated by:

- A rise in cytosolic Ca^{2+}
- Inositol triphosphate
- cAMP.

T_3 and T_4 may regulate TSH secretion by feedback to the hypothalamus. TRH can stimulate the release of prolactin.

Corticotrophin-releasing hormone (CRH)

Corticotrophin-releasing hormone (CRH) activates the release of adrenocorticotrophic hormone (ACTH). ACTH acts on the adrenal cortex where it causes the release of cortisol and corticosterone. CRF release is controlled by higher nervous centres and release is affected by:

- Stress
- Hypoglycaemia
- Trauma and surgery
- Fever.

There is a circadian rhythm, which is highest at 09.00 h and lowest at about midnight. This rhythm is lost in response to stress. The period of the rhythm is established at 3 years of age and is present for life. The rhythm is also controlled by the pattern of sleep. This involves a negative feedback control by cortisol. Decreased CRH production leads to:

- Increase in plasma ACTH levels ('short feedback loop').
- Increase in plasma cortisol levels ('long feedback loop').

Gonadotrophic-hormone-releasing hormone (GnRH) (in older terminology: luteinizing hormone-releasing hormone and follicle-stimulating releasing hormone)

The gonadotrophic hormones – follicle-stimulating hormone (FSH) and luteinizing hormone (LH) – are stimulated by GnRH. GnRH is produced by the basal medial area of the hypothalamus and its release is pulsatile with a periodicity of 1–2 h.

Prolactin-releasing hormone (PRH) and prolactin release inhibitory hormone (PIH)

Secretion of prolactin is pulsatile, with higher levels being produced at night. The physiological causes of increased prolactin levels include:

- Pregnancy
- Suckling and nipple stimulation
- Coitus
- Sleep
- Stress
- Puberty in girls.

PIH is now known to be a dopamine and it exerts an inhibitory control of prolactin release. Other inhibitors are:

- Catecholamines
- Histamine.

Putative prolactin-releasing hormones are:

- 5-HT
- Angiotensin II
- Vasoactive intestinal polypeptide.

TRH stimulates the release of prolactin.

HORMONES OF THE ANTERIOR PITUITARY

Growth hormone (GH)

The major function of growth hormone is growth. It stimulates the growth of bones and other tissues associated with a number of other factors:

- Adequate nutrition.
- Other hormones:
    Sex hormones
    Thyroid hormones
    Insulin.

GH causes the growth of most tissues, bringing about an increase in cell size. There is also an increase in cell numbers (mitosis). GH is also involved in the differentiation of some cells, e.g. cartilage and early muscle cells. It is important during special periods in the growth of an individual. This is particularly the case during accelerated growth at 2 years and at puberty. At puberty it causes an increased level of sex hormones and adrenal hormones.

Cessation of growth

Sex hormones are also associated with the fusion of the growing ends of bones, epiphyseal closure. Bone growth is then complete. Various problems are associated with GH and epiphyseal closure.

Before closure

Before closure excess GH secretion produces giantism where parts of the body grow in proportion. It is not uncommon for individuals to grow to heights of 7–8 feet (2.1–2.4 m).

After closure
After closure excess GH secretion results in acromegaly. This condition arises from an acidophil tumour of the pituitary, resulting in continuous hypersecretion (no physiological control). There is a resulting overgrowth of membrane bone and soft tissues, especially of the hands, feet and face. Diabetes may develop.

Deficiency of GH in children results in dwarfism. In this condition:
- Growth rate is reduced.
- The children are small for their age.
- They are also sexually underdeveloped.

Effects on metabolism
GH increases amino acid transport across cell membranes and their incorporation into proteins. There is also an increased translation of messenger RNA to protein and increased transcription of DNA in the nucleus, causing an increased cellular messenger RNA content.

GH also influences carbohydrate and fat metabolism. Its effects are, in general, opposite to those of insulin, an increase in blood glucose (diabetogenic action). GH stimulates lipolysis in adipose tissue and thereby increases fatty acid release. It further enhances oxidation of fatty acids and promotes the use of fat rather than glucose.

Mediation of the action of GH
Apart from the action on amino acids which appears to be direct, GH exerts its metabolic effects through substances (polypeptides) known as *insulin-like growth factors (IGF)*. Two of these are known as IGF-I and IGF-II and their major site of synthesis is the liver.

Corticotrophin (ACTH)
Adrenocorticotrophic hormone is a polypeptide of 39 amino acids. It acts on the adrenal cortex. ACTH binds to specific membrane receptors on the adrenal cortical cells. There it activates adenylate cyclase, resulting in an increase in cellular cAMP. This increases the synthesis of steroids and stimulates the secretion of glucocorticoids. These hormones in turn affect the metabolism of carbohydrates, proteins and fats.

Actions of glucocorticoids

Effects on intermediary metabolism
Stimulate glycogenesis and glycogen storage.
Stimulate gluconeogenesis during fasting→glucose for brain.
Raise blood sugar (counter-insulin) by inhibiting glucose uptake by muscle and adipose tissue.
Promote fatty acid mobilisation.

Effects on circulatory function
Help in the maintenance of myocardial contractility.
Potentiate vasoconstrictor action (e.g. noradrenaline).
Decrease vascular permeability.

Reaction to stress
Cortisol secretion increases during stress.
In large doses glucocorticoids:
Suppress the immune response, e.g. to prevent transplant rejection
Are effective in treating inflammatory and allergic responses.

Prolactin

Prolactin is necessary for breast development. It is also essential in the initiation and maintenance of lactation. Development of the alveoli and ducts requires several hormones, cortisol, chorionic somatomammotrophin and growth hormone, as well as oestrogen and progesterone. Although oestrogen and progesterone are necessary for breast development, they inhibit milk secretion. The concentrations of blood prolactin of pituitary origin increase about 10 times from the 5th week of pregnancy to parturition. The placenta produces chorionic somatomammotrophin – it acts as a lactogen. At parturition hormone levels of oestrogen and progesterone fall. This allows prolactin to initiate milk secretion, which also requires GH, cortisol and parathyroid hormone. Prolactin levels fall after birth.

Suckling causes tactile stimulation of the nipples. This activates afferent nervous impulses which results in a surge of prolactin release. This release maintains secretory function of the breasts. Suckling also releases oxytocin. This initiates contractions in the myoepithelium and subsequently the ejection of milk (the milk let down). Prolactin exerts an inhibitory effect on the ovaries. In consequence:

- Oestradiol and progesterone production are reduced.
- Conception is prevented.

THE THYROID GLAND

HORMONES OF THE THYROID GLAND

The hormones of the thyroid gland are:
- Thyroxine – T_4.
- Tri-iodothyronine – T_3.

Activity of the thyroid gland is regulated by thyroid-stimulating hormone (TSH) from the anterior pituitary, where it:

- Releases thyroid hormones.

- Increases iodine uptake from the plasma.
- Promotes hyperplasia of the gland.

SYNTHESIS OF THYROID HORMONES

Dietary iodine is reduced to iodide in the small intestine before absorption into the blood. Inorganic iodide is taken up from the plasma by the thyroid gland. There is an iodide pump which can concentrate iodide from 30- to 250-fold. Inside the thyroid cell iodide is oxidised to iodine by H_2O_2 (with iodine peroxidase). Thyroglobulin (TGB) is synthesised on the rough endoplasmic reticulum. Iodine combines with the side chains of tyrosine in the TGB molecule to form monoiodotyrosine (MIT).

$$\text{MIT} + \text{iodine} \xrightarrow{\quad \text{iodinase} \quad} \text{di-iodotyrosine (DIT)}$$

MIT + DIT and DIT + DIT occurs on the surface of TGB to form T_3 and T_4. The iodinated TGB passes into the lumen of the follicle where it is stored as colloid.

For secretion of T_3 and T_4 to occur, TGB must be broken down. Drops of colloid pass into the acinar cells by pinocytosis which then fuse with lysosomes. The proteolytic enzymes of the lysosomes break down TGB, producing T_3, T_4, MIT and DIT. T_3 and T_4 pass into the circulation.

The iodotyrosines (MIT and DIT) are de-iodinated and the iodine recycled within the cell. The thyroid secretes about 100 nmol (80 μg) of T_4 and 8 nmol (5 μg) of T_3 a day. It is detoxified in the liver and secreted in the bile as glucuronides or sulphates.

Daily requirements of iodine
250 μg/day.

Plasma levels
T_4 75–140 nmol/L.
T_3 1.1–2.3 nmol/L.

Thyroid hormone transport
Ninety-nine per cent of T_3 and T_4 are bound to plasma proteins:

- Thyroxine binding globulin – binds 75–80% of both T_3 and T_4.
- Thyroxine binding pre-albumin – binds 10–15% of T_4 and 10% of T_3.
- Albumin – binds smaller amounts with low affinity.

It is thought that T_4 exerts most of its effects through conversion to T_3.

THYROID HORMONE ACTIONS

Uptake by cells

T_4 (0.02%) and T_3 (0.3%) free in the circulation are considered to be the active hormones. Cells also have specific receptors for these hormones, which enter cells by active transport. Within the cell transport to the nucleus may be through specific transport proteins. The movement of T_3 into the nucleus may be stereospecific and energy-dependent in liver, anterior pituitary, kidney and other tissues. It is essential for normal growth and development and it also influences the basal metabolic rate.

General metabolic effects
Thyroid hormones stimulate nuclear transcription of a large number of genes; this causes an increase in enzymes and transport proteins. The metabolic activities of almost all tissues are affected.

Protein metabolism
Increased amino acid uptake and protein turnover lead to an increase in both synthesis and catabolism. Many aspects of carbohydrate metabolism are affected:

Glucose uptake.
Glycolysis and gluconeogenesis.
Increased insulin secretion→secondary consequences.

Fat metabolism
Increased fat mobilisation.
Increased fatty acid concentrations.
Increased fat oxidation.

Iodine deficiency
Iodine is essential for synthesis of thyroid hormones. Deficiency leads to:

• Increased TSH production.
• Enlargement of thyroid gland (goitre).

DISEASES OF THE THYROID GLAND
Thyroid disease results in:

In the adult:
Hypothyroidism (myxoedema).
Fall in metabolic rate.
Subnormal temperature.
Patient feels the cold.
Slow pulse.
Deposits of proteoglycan→puffiness of the hand and face, thickening of skin.
Weight increases.
Mental processes retarded.
Disturbances of sexual function.

In the child:
Cretinism.
Child is small.
Mentally defective.
Coarse scanty hair.
Thick scaly skin.

Hyperthyroidism (thyrotoxicosis)

Hyperthyroidism results in an increased metabolic rate. The indications are a flushed skin with:

- Vasodilatation
- Sweating
- Increased heat loss.
  The patient does not feel the cold and has a rapid heart rate. There is often weight loss as a consequence of:

- Catabolism of protein
- Oxidation of fats.
  Additional actions include:

- Increased motility and secretion of the gastrointestinal tract.
- Breathlessness on exertion.
- Nervous excitability→rapid fine tremor.
- Enlargement of thyroid→goitre.
- Protrusion of eyeballs→exophthalmos.
- Disturbances of sexual function.

Goitrogens or antithyroid substances

Block of iodine uptake:

- Thiocyanate
- Perchlorates.
  Block of hormone synthesis:
- Thiurea
- Imidazole.

Calcitonin

Calcitonin is a hormone secreted from the thyroid.

Site of origin
The parafollicular or C cells lying adjacent to the follicles.

Function
In calcium homeostasis.

THE MALE REPRODUCTIVE SYSTEM – THE TESTES

The testes are a pair of organs located in a sack, the scrotum. The functions of the testes include:

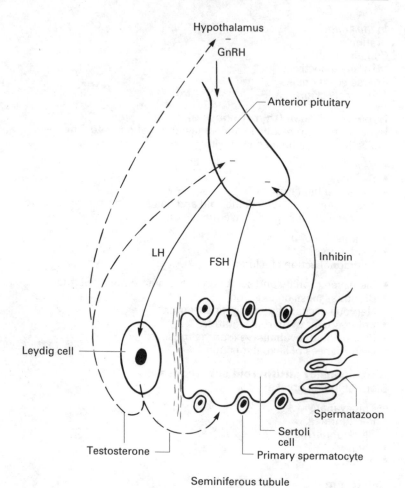

Fig. 8.2
Control of spermatogenesis and testosterone secretion.

- Production of spermatozoa.
- Secretion of male steroid hormones.

SPERMATOGENESIS (Fig. 8.2)

At about 8 months of uterine life the testes descend into the scrotum. Spermatogenesis only occurs at temperatures 2–4°C below the abdominal temperature. Temperature control is brought about by:

- Contraction of dartos and cremasteric muscle – contract in the cold, relax in the warmth (regulates the distance of testes from the groin).
- Blood flowing to the testis is cooled by a counter-current flow in the pampiniform plexus of veins.

An adult testis contains about 750 seminiferous tubules (70 cm long and 200 μm wide). Tubules have a basement membrane on which there are sets of cells made up of two types:

- Spermatogenetic
- Supportive.

The outermost layer (cubical) of the testis is the *spermatogonia*.

Both FSH and progesterone are necessary for spermatogenesis. The time required for spermatogenesis is about 70 days. The spermatozoa are released into seminiferous tubules and are non-motile. As the spermatozoa pass into the epididymis and mature, they are stored until ejaculated. The production is a continuous process and those not ejaculated deteriorate and are phagocytosed. Spermatozoa are first produced at puberty and continue until old age.

Supportive cells

Supportive cells provide germinal cells with nutrients and stimulating factors. They secrete an androgen-binding protein and the hormone *inhibin*. There are two types of cell:

1. Sustentacular cells of Sertoli.
2. Non-germinal cells.

Mature sperm

The mature sperm is a package of paternal genes powered by an active tail. The sperm head contains the male pronucleus and is covered by a cap, the *acrosome*. The tail (organ of motility) consists of the neck, middle piece, main piece and end piece. The middle piece contains mitochondria which provide the energy for motility. The sperm obtain energy from nutrients secreted by the part of the genital tract they are traversing. As many as 30% may be abnormal: e.g., decapitated; double headed or tailed; multinucleate giants.

SEMEN

Semen is the fluid ejaculated during the male orgasm. It functions to carry spermatozoa. The fluid contains fructose and citrate which are a source of energy for spermatozoa in the presence of O_2. It also contains prostaglandins which may increase the motility of the uterus following intercourse. Fibrinolysin liquefies semen which clots after ejaculation. It also contains hyaluronidase, which acts on the cervical mucus to weaken it and allow sperm to pass through the cervix. The fluid from the bulbo-urethral glands act as a lubricant. The normal volume of normal ejaculate is 2–5 ml. It is made up of:

- About 10% spermatozoa
- 90% seminal plasma.

Ejaculate consists of secretions from the:

- Epididymis
- Seminal vesicles
- Prostate
- Bulbo-urethral glands.

Composition
The seminal fluid is derived from:

Seminal vesicles
The seminal vesicles contribute about 70% of seminal plasma at a pH = 7.4. The fluid contains:

- Fructose
- Ascorbic acid
- Prostaglandins
- Hyaluronidase.

Prostate
The prostate contributes about 20% of the seminal fluid. It is colourless and slightly acid (pH 6.5) due to citric acid. It is composed of:

- Albumin and proteolytic enzymes fibrinolysin, fibrinogenase
- Acid phosphatase
- Antibacterial substance.

PUBERTY

In the male puberty begins at 10–14 years of age. It is associated with accelerated body growth and development of secondary sexual characteristics. The first external signs of growth are the penis and testes. This is soon accompanied by the appearance of pubic hair and later the appearance of axillary and facial hair. In the middle of pubescence the voice deepens, the vocal cords lengthen and spermatogenesis begins. Testosterone secretion increases about 60-fold.

TESTOSTERONE

Testosterone secretion begins in the fetus where it plays an essential role in the differentiation of the male reproductive tract and brain. At puberty it stimulates the growth and functional maturity of reproductive tract and brain. Testosterone also sustains spermatogenesis. At puberty it is responsible for the secondary sexual characteristics. It influences the anterior pituitary and, by negative feedback, inhibits the release of LH. This stimulates the

Sertoli cells to produce androgen-binding protein which is produced by the interstitial cells of Leydig (cells lying between the seminiferous tubules).

Testosterone is released into the blood where it is bound to sex hormone-binding globulin. In target tissues (except muscle) testosterone is converted to *dihydroxytestosterone*. This complex later combines with nuclear receptors capable of gene activation resulting in protein synthesis. In the liver it is conjugated to glucuronides and excreted by the kidney.

CONTROL OF MALE REPRODUCTIVE ACTIVITY

Spermatogenesis depends upon:

- FSH
- LH
- Testosterone.

Anterior pituitary gonadotrophic hormones (FSH and LH)

FSH controls spermatogenesis (development beyond the primary spermatocyte requires testosterone).
LH controls the synthesis of testosterone.
Release of gonadotrophic hormones is pulsatile, controlled by GnRH.
Testosterone inhibits the release of LH by negative feedback on the hypothalamus.
FSH output is inhibited by *inhibin* secreted by Sertoli cells.

Other sites of androgen production

Androgens are produced in both the male and female in the adrenal cortex:
Testosterone
  Androsterone
  Dehydroepiandrosterone.

Possible sites of oestrogen production

Interstitial cells of Sertoli are a site of production of oestrogen. Tumours of these cells can lead to feminization of the male. In this condition the adrenal cortex produces oestrogens.

Bilateral vasectomy

Bilateral vasectomy is a common elective surgical procedure to remove the capacity to conceive. The procedure results in:

- No loss of testicular weight.
- No change in histology.
- Continued spermatogenesis.
- Phagocytosis of sperm in epididymis.
- Prostate and seminal vesicles unaffected.
- No change in volume of semen.

THE FEMALE REPRODUCTIVE SYSTEM – THE OVARIES

The primary female reproductive organs are the ovaries. There are two ovaries which:
- Produce the ova (gametes).
- Secrete the sex hormones – *oestrogen* and *progesterone*.

Both functions are regulated by the hypothalamus and adenohypophysis.

ACCESSORY REPRODUCTIVE STRUCTURES

There are two fallopian tubes (uterine tubes), uterus, cervix, vagina and external genitalia (vulva). The control and integration of the reproductive system can be considered at four levels:
- Hypothalamus
- Adenohypophysis
- Ovaries
- Uterus.

OVARIES

The ovaries are two flattened bean-shaped structures 3–4 cm in length, 2 cm in breadth and 1 cm thick. They are covered by a germinal epithelium which lies over a layer of connective tissue, the tunica albuginea. The eggs (oogonia) occur in the germinal epithelium but no oogonia are formed after fetal life. The normal female has about 750 000 follicles at birth.

The oogonia produce the primary oocytes which are arranged in primary follicles. These are surrounded by a single layer of cells, the *granulosa*, which in its turn is surrounded by the *theca interna*, a layer of stromal cells. At the beginning of an ovarian cycle several follicles begin to enlarge, but only one continues, the others regress – a process called *atresia*. The granulosa and thecal cells proliferate and a fluid is secreted by the granulosa cells, causing a space to occur in the follicle. This displaces the oocyte to one side where it is surrounded by granulosa cells – the *cumulus oophorus*. Between the oocyte and the granulosa cells is the *zona pellucida*. Processes from both the granulosa cells and the oocyte penetrate into the zona pellucida and exchange nutrients. The theca interna cells enlarge and become vascularised. Further enlargement of the follicle occurs, compressing the stroma, this causes the theca externa to form a false capsule – this is the primordial follicle. The primordial follicle is typically 30 μm in diameter; the mature follicle can be as large as 10–30 mm.

OVULATION

The mature follicle protrudes from the surface of the ovary. It then ruptures and discharges its contents into the peritoneal cavity – the ova with its zona pellucida, corona radiata and cumulus oophorus still attached.

The fimbria of the uterine tubes move to the surface of the ovary and facilitate the entry of the oocyte into its abdominal opening. The oocyte is rapidly carried to the ampulla (where it is fertilised) by peristalsis of the tubes and the ciliary activity of the tubal epithelium. In women, ovulation is induced spontaneously and not by coitus.

Time of ovulation
Ova can be found in the fallopian tube by washings by about the 15th day of the menstrual cycle. Ultrasound scans reveal rupture at about the 14th day. The follicle is about 2 cm in diameter and contains about 10 ml of fluid.

Guides to time of ovulation
Several indicators can be used as indicators of ovulation:
- As a guide to ovulation body temperature changes. This involves a fall, followed by a small rise (0.5°C) in the middle of the menstrual cycle.
- Plasma and urine oestrogen levels are seen to change 24 h before a rise in LH. This is followed by ovulation 24–36 h later.
- Plasma progesterone of more than 13 nmol/L by day 21 of the cycle is good evidence that ovulation has occurred.

Ovarian events after ovulation
At ovulation changes occur at the site of the follicle.

Luteinisation
The follicle collapses and the granulosa cells left proliferate rapidly and become vascularised. These cells form the bulk of the growing *corpus luteum* – it is a yellow colour due to lipids in the cells. The corpus luteum continues to secrete *oestrogen* and also produces *progesterone*. It is functional for about 12 days after which it regresses. After a few months a structureless white mass is left – the *corpus albicans*.

THE GONADOTROPHIC HORMONES

In order to mature, the follicles require both FSH and LH from the anterior pituitary. The secretion of FSH and LH is stimulated by GnRH, which comes from the hypothalamus. The release of GnRH is pulsatile, with a periodicity of 1–2 h. Secretion of FSH and LH are regulated by oestrogen, progesterone and inhibin (see later). FSH causes growth and development of the follicle, but does not induce ovulation. Oestrogens are secreted by the granulosa cells under FSH control. LH acting together with FSH induces ovulation and the formation of the corpus luteum. FSH together with oestrogen

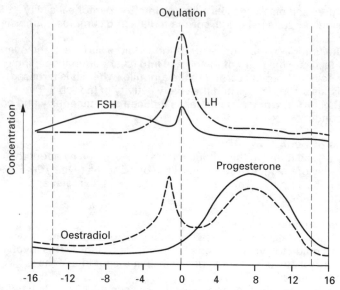

Fig. 8.3
Hormones during the ovarian cycle.

causes the development of the follicle. LH is involved in the
secretion of oestrogen by the ovary.

OVARIAN HORMONES (Fig. 8.3)

The ovary secretes three hormones:
- Two steroids:
    Oestrogen – oestradiol-17β is the principal hormone
    Progesterone.
- One protein:
    Inhibin.

Oestrogen is secreted by the granulosa cells and by the corpus
luteum. In the first part of the follicular phase oestrogen secretion
increases very little. However, as the follicle grows oestrogen
secretion rises which stimulates LH secretion from the pituitary. The
high level of oestrogen exerts a strong positive feedback and
triggers a discharge of LH and FSH (greater effect on LH), which
lasts about 24 h. As FSH and LH rise there is a concomitant increase
in secretion of progesterone and a steep fall in oestrogen. The
progesterone level remains elevated for most of the luteal phase
and then falls after FSH and LH have returned to basal levels.
Oestrogen and progesterone act synergistically to exert a negative
feedback on the secretion of FSH and LH. The feedback systems are
associated with:

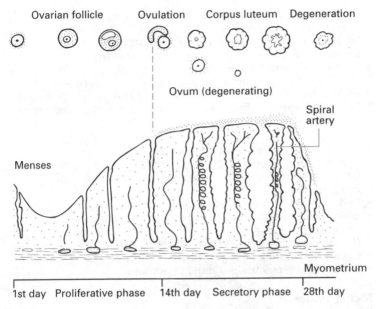

Fig. 8.4
The menstrual cycle: changes in the endometrium, the ovarian follicle and the corpus luteum.

- Negative feedback of oestrogen exerted on the pituitary.
- Positive feedback of oestrogen exerted on the hypothalamus.
- Negative feedback which diminishes as the corpus luteum regresses.

THE MENSTRUAL CYCLE (Fig. 8.4)

The menstrual cycle is a cycle of hormonal and structural events. From puberty, the endometrium of the uterus undergoes cyclical changes that result in bleeding or menstruation lasting 4–6 days. This recurs about every 28 days if an oocyte is not fertilised. In the menstrual cycle events are dated either from the first day of bleeding or from the peak of plasma LH. The latter is then termed day 0 (zero).

The cycle is divided into two phases:
1. The follicular or proliferative phase.
2. The luteal or secretory phase.

Follicular or proliferative phase (first half of cycle)
Follicular changes refer to the ovary and proliferative phases refer to the uterus. The following sequence occurs:
- Ovarian follicle enlarges with an increased oestrogen secretion.

- Oestrogen acts on the uterine endometrium to stimulate regeneration and growth.
- On the 5th to 14th days new capillaries grow from spiral arteries.
- Endometrial glands grow in length and tortuosity.
- Cells of the stroma hypertrophy.
- Height of the endometrium increases two to three times.
- Follicle comes to the surface of the ovary and ovulation occurs.
- Corpus luteum develops and both oestrogen and progesterone are secreted.

Luteal or secretory stage (second half of cycle)
The sequence of events in this stage of the cycle are:
- Corpus luteum develops and secretes both oestrogen and progesterone.
- Endometrium is prepared to receive a fertilised ovum.
- Glands continue to grow and become more dilated and convoluted.
- Glands become distended with secretory products.
- Cells of the stroma enlarge and accumulate glycogen.
- Endometrium reaches 6–7 mm in thickness.

Menstruation
If the ovum is not fertilised, then menstruation occurs. The corpus luteum regresses which brings about a rapid fall in plasma oestrogen and progesterone. The blood flow through spiral arteries slows as menstruation approaches. Then there is an intense vasoconstriction of the arteries which stops blood flow to the endometrium for some hours. As a consequence there is a blanching of endometrium followed by degeneration. The arteries then open up and blood escapes. This is called a subepithelial haematoma. Blood clots can form but the fibrinolytic system is simultaneously activated. The clots are broken down by plasmin allowing the menstrual flow to proceed.

Blood loss
Eighty per cent of women lose 6–60 ml of blood. This contains:
- 3–30 mg iron
- 1–12 g protein.

Fertilisation
Fertilisation depends critically on the viability of spermatozoa and ovum.

Viability of spermatozoa
Microscopic examination of spermatozoa is of limited value but abnormally shaped sperm are unlikely to fertilise. Motility can

persist even after the power to fertilise is lost. The ability of sperm to fertilise lasts about 24 h.

Viability of ova
The ovum is viable for a shorter time than the sperm, about 8 h. The unfertilised ovum degenerates in the uterine tubes.

Male ejaculate
The male ejaculate (see the section on semen, p. 267) contains about 100 million spermatozoa per millilitre. The process of ejaculation involves a reflex action and emission takes place in an ordered sequence:
- During erection the bulbo-urethral glands discharge. This aids lubrication.
- During ejaculation parts of the semen appear in the following order:
  Prostatic secretion which is alkaline→neutralises acidity of male urethra and vagina;
  Spermatozoa;
  Seminal vesicle secretion.
- Spermatozoa must remain in the genital tract some hours before being capable of fertilising an ovum where a process of *capacitation* occurs. This involves:
  Breakdown and merging of the plasma membrane and acrosomal membrane;
  Release of enzymes thought to play a role in sperm penetration.
- Cumulus cells surrounding the ovum are dispersed by bicarbonate and hyaluronidase.
- Sperm penetrate zona pellucida by enzyme action (acrosomal proteinase).
- Sperm enters ovum – only one out of the whole ejaculate.
- Zona pellucida becomes impenetrable by other spermatozoa.

PREGNANCY

Fertilisation usually occurs in the uterine tube within 12 h of coitus. The ovum is transported down the uterine tube by ciliary action and peristaltic movements. After fertilisation the zygote begins to divide on its way to the uterus (about 3 days). Cell division continues to the *morula* stage (64–128 cells). The morula then develops into a blastocyst which has an outer layer of trophoblastic cells, forming a fluid-filled cavity in which the inner mass of embryonic cells develops. The blastocyst implants into the endometrium. The trophoblastic layer forms part of the chorion – the fetal part of the placenta. This layer secretes *chorionic gonadotrophin* which has an LH effect maintaining the corpus luteum. Progesterone secretion thus continues and transforms endometrium into a *decidua*. The decidua consists of large polyhedral cells laden with glycogen. For the next 6–8 weeks the corpus luteum continues to be the main source of progesterone. After 6–8 weeks the placenta takes over as the main source of progesterone.

FORMATION OF THE PLACENTA

The placenta develops from the trophoblast cells. The endometrium continues to grow and reaches 10 mm thick. The superficial layer of stroma becomes more compact and the cells enlarge to become decidual cells. In the decidua (basalis) under the embryo the maternal blood vessels dilate. Some small arteries of the decidua necrose and spaces are formed full of maternal blood. Finger outgrowths from the outer layer of the blastocyst and the *chorionic villi* grow into these spaces. The villi are invaded by mesoderm carrying fetal blood vessels. This allows the fetal and maternal circulations to come in close contact. There is no transfer of blood between mother and fetus.

IMMUNOLOGICAL STATUS OF THE FETUS

The fetus has antigens foreign to the mother (genes derived from the father). Gestation is longer than a homograft rejection period and rejection is a possibility. This is prevented and the fetus obtains protection from:
• Placenta.
• Immunosuppressive substances in mother's blood?
• Progesterone is essential for the maintenance of pregnancy, possibly due to its ability to inhibit T-lymphocyte cell-mediated responses involved in graft rejection.
  Placental protection may not be complete:
• Red cells may cross.
• Rh antibody may pass back into the fetus.
  See Chapter 1.

AMNIOTIC FLUID

Amniotic fluid is secreted by the amniotic epithelium. It also contains fetal urine and lung liquid. The volume changes with fetal development:
• 30 ml at 10 weeks.
• 350 ml at 20 weeks.
• 500–1100 ml, maximal volume, at 35 weeks.
  Its function is to maintain the fetus in a safe environment, warm,

'weightless' and shockproof. The fetus can swallow amniotic fluid and passes urine into it. The amniotic fluid contains fetal cells which can be harvested by amniocentesis, and used for diagnostic purposes:

- Determination of sex.
- Fetal abnormalities.
- Down's syndrome.

HORMONAL CONTROL OF PREGNANCY

The ovary is necessary in the early stages of pregnancy. The corpus luteum secretes progesterone which prepares the endometrium. The placenta provides all the hormones necessary in the later stages of pregnancy.

Four hormones are secreted by the placenta

   Protein hormones
Chorionic gonadotrophin
Chorionic somatomammotrophin.
   Steroid hormones
Oestrogen
Progesterone.
   Other important hormones are:

Chorionic gonadotrophin (CG)
CG is a glycoprotein secreted by trophoblastic cells. It is chemically and biologically similar to LH. There are high levels in the first trimester of pregnancy, which are responsible for maintaining the corpus luteum. The corpus luteum secretes:

- Oestradiol
- Progesterone.

Chorionic somatomammotrophin (CS) (placental lactogen)
CS is structurally similar to growth hormone but has weak growth-stimulating properties when compared with GH. Its concentration in maternal blood increases steadily throughout pregnancy. CS is secreted primarily into the maternal circulation but is found in low levels in cord blood. It promotes mammary development and may be responsible for some metabolic changes seen in pregnancy:

- Increased secretion of insulin.
- Increased insulin resistance.
- Increased plasma fatty acid levels.

Oestrogens
The main steroid secreted by the placenta is oestriol. The placenta cannot form 17-hydroxyprogesterone, its precursor. The precursors

of 17-hydroxyprogesterone are synthesised by the fetus and maternal adrenal glands.

Feto-placental unit – production of progesterone and oestradiol

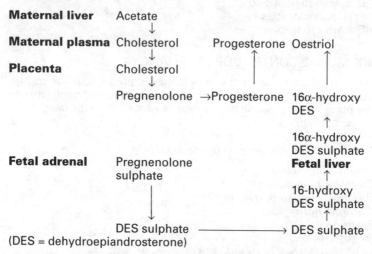

Maternal liver	Acetate
Maternal plasma	Cholesterol	Progesterone	Oestriol
Placenta	Cholesterol

Pregnenolone →Progesterone 16α-hydroxy DES

16α-hydroxy DES sulphate **Fetal liver**

Fetal adrenal Pregnenolone sulphate

16-hydroxy DES sulphate

DES sulphate ⟶ DES sulphate
(DES = dehydroepiandrosterone)

Plasma levels rise steadily during pregnancy but levels of free oestrogen are not increased so markedly because of the concomitant increase in the sex hormone-binding globulin.

Oestrogen is necessary for uterine and mammary gland development.

Progesterone
Levels of progesterone rise 3 weeks after fertilisation. They then fall, only to rise again after about 8 weeks until the end of pregnancy and parturition. Progesterone remains high during labour and falls towards the end of parturition. It is needed to maintain the endometrium and to suppress spontaneous contractions of the myometrium. An important function is inhibition of prostaglandin formation. It is also involved in blocking the cellular immune response to protect the placenta.

FUNCTIONS OF THE PLACENTA

The placenta increases in weight throughout pregnancy to about 500 g at term. It has two main functions:
• Mediates transfer between mother and fetus.
• As an endocrine organ.

Maternal and fetal vascular networks
The maternal and fetal vascular networks are separated by 2–3 μm and have an area of 12 m. The general structure:

- Maternal side – sinusoids.
- Fetal side – chorionic villi capillaries.

Fetal–maternal barrier – the chorion and fetal capillary wall

The chorion has properties of an epithelium. It is permeable to lipid-soluble substances and restricts the passage of polar compounds. It has active transport mechanisms to transport specific substances.

Transfer of nutrients

The fetal capillary bed provides a very large surface area in contact with maternal blood.

Oxygen

Transported by simple diffusion.

A full-term fetus requires 15–25 ml O_2/min.

The A-V difference in supply to uterus 7 ml/100 ml.

Fetal requirements are met with a blood flow of 200–300 ml/min.

Uterine blood flow is 500–750 ml/min.

Partial pressure of oxygen supplying the fetus is low (20–25 mmHg).

To achieve a large O_2 carrying capacity at a low Po_2:

  The fetus is polycythaemic and has $6–8 \times 10^{12}$ cell/L

  A haemoglobin concentration of 14–22 g/L

  An oxygen dissociation curve displaced to the left –

    Higher affinity for O_2 by fetal Hb, high proportion of HbF

    Low concentration of 2,3-BPG (DPG).

Lipids

Lipid-soluble substances cross the placenta readily. If the mother indulges in narcotics, alcohol or nicotine, these have easy access to the fetal circulation and may result in low birth weight (smoking), fetal malformations (excess alcohol) and withdrawal symptoms (narcotics).

Glucose

This is carried across the placenta on specific carrier mechanisms (see p. 135).

Proteins

As with glucose, there are specific carrier mechanisms for amino acids. The fetus receives about 1 g/kg nitrogen from amino acids per day: two-thirds is used for growth; one-third is used for energy, the nitrogen is returned to the mother as urea.

Maternal hyperphenylalaninaemia (phenylketonuria)→high fetal blood levels→damage to fetal brain.

Electrolytes and water
Most minerals appear to reach the fetus by carrier-mediated transport. Although transferrin does not cross the placenta, iron reaches the fetus in some bound form.

Antibodies
IgG immuoglobulins cross by a specific transport mechanism. IgA and IgM antibodies are synthesised by the fetus.

PHYSIOLOGICAL RESPONSES OF THE MOTHER TO PREGNANCY

Body fluids and blood
Total body water is increased as pregnancy continues. This is due to an increased concentration of oestrogens.

The constituents of blood are also altered:
- Red cell mass increases.
- Plasma volumes increase.
- PCV and HB fall.
- β-Globulin and fibrinogen rise.

	Pre-pregnancy	**38th week of pregnancy**
Plasma volume (ml)	2297	3615
Red cell volume (ml)	1119	1509
Red cell count	4.35×10^6/L	3.85×10^6/L
Haemoglobin (g/dl)	13.4	11.0
Haematocrit (PCV)%	38.75	33.12

Mean figures from 36 women not on iron supplementation. From Lind T. (1985) Maternal Physiology

Cardiovascular functions
Heart increases in size (about 12% from radiology).
Cardiac output increases from a mean of 4.88 L to 7.2 L.
Systolic blood pressure is relatively little changed, diastolic pressure falls.

Kidney
GFR increases 97–128 ml/min (by 10th week).
Effective renal plasma flow increased 480–890 ml/min.
Urinary output/24 h largely unchanged.
Glucose and amino acids may appear in the urine.

Weight
Total gain about 12.5 kg.
Greater than expected from water retention:
  Depot fat 3.5 kg
  Protein 0.8 kg (mainly uterus and content).
  Monitoring of weight is important:

Sudden increase in late pregnancy – sign of developing pre-eclampsia.
Failure to gain weight:
   Placental insufficiency
   Poor fetal growth.

Energy requirements
Increased tissue stores account for 170 MJ.
Cumulative extra needs of fetus and uterus 150 MJ.
Supplements of iron to combat anaemia.
Supplement of folic acid to prevent neural tube defects.

Endocrine function
Renin–angiotensin system:
   Renin levels increased
   Levels of angiotensinogen increased
   Angiotensin I and II also increased
   Vascular response to angiotensin II decreased.
Placental hormone production (see above).
Hypophysis enlarges.
Increased secretion of trophic hormones.
Melanocyte-stimulating hormone – may be responsible for pigmentation.
Corticosteroid increase – bound to transcortin.
Thyroid enlarges.
Raised plasma-bound iodine (PBI).
Reduced tolerance of glucose – in spite of increased circulating insulin.

PARTURITION

Duration of pregnancy
The duration of pregnancy can vary. Labour occurs:
- At 280 ± 7 days in 55%.
- Before 273 days in 25%.
- After 287 days in 20%.

   Uterine contractions occur throughout pregnancy, but there is an increase in amplitude and frequency towards the end of pregnancy. The mechanisms which initiate parturition are still unknown. It is easier to say what is not involved than to give a definite explanation. The hormone relaxin is secreted towards the end of pregnancy.

Relaxin
Relaxin is produced by the ovaries and prepares the birth canal by:
- Softening the cervix.
- Softening the cartilage of the symphysis pubis.

Oestrogen and progesterone
Oestrogen and progesterone levels continue to rise towards
parturition. Their fall afterwards is regarded as the result and not
the cause of parturition.

Oxytocin
As the pregnant uterus is sensitive to oxytocin, it has been
suggested that oxytocin is responsible for the initiation of
parturition. This is based on the following observations:
- Intravenous infusions of oxytocin can bring on parturition.
- Hypophysectomy is without effect.
- Plasma oxytocin levels are low in pregnancy and do not rise
  before the onset of labour.
- Its role may be in the expulsive phase of labour and in the
  reduction of blood loss following delivery.

Prostaglandins
Prostaglandins administered to a pregnant women will induce
uterine contractions.
Prostaglandin inhibitors will prevent or avert preterm labour.

Fetal triggering
It has been suggested that signals originating from the fetus initiate
parturition. Evidence that the fetal adrenal may play a part has not
been substantiated. Prostaglandins have been implicated. The
trigger which releases prostaglandins is not known, but it is
suggested that they arise in the fetus and exert their effects on the
uterus through a paracrine mechanism.

Once labour is initiated, stimuli from the genital tract cause reflex
release of oxytocin.

Labour
Labour can be divided into three stages:

First stage
Uterine contractions begin and intrauterine pressure increases with
each contraction: 30–50 mmHg above basal (1–5 mmHg).
Pain occurs with each contraction.
Uterine contractions dilate the upper part of the cervical canal and
later the os uteri.
Membranes rupture.

Second stage
Intrauterine pressure reaches 110 mmHg.
Contractions more frequent.
Head of child gradually forced through pelvis and is born.
Voluntary increases in intra-abdominal pressure cause uterine
pressure to reach 260 mmHg.

Third stage
After the birth of the child – uterine contractions cease for 5–10 min.
Contractions start again – placenta and membranes are expelled.
Detachment of the placenta leaves a raw bleeding place.
Average loss of blood 300 ml.

Puerperium
This is the period following parturition. During this time the uterus
involutes.

THE FETUS

The fertilised egg is 80 μm in diameter, while the fetus at delivery is
50 cm in length and weighs 3.4 kg. Growth is rapid at first but slows
near term. Nearly half the weight is acquired is in the final 6–8
weeks. The weight of the newborn drops immediately after
delivery.
　The maturity of the fetus is estimated by measuring the rate of
growth by:
• Clinical assessment
• Ultrasound.
　The fetus draws on the mother's tissues if the mother's nutritional
intake is inadequate. Birth weight is not governed by maternal diet,
unless there is malnutrition.

MEAN BIRTH WEIGHTS

The following are typical values:

Singletons	3.38 kg
Twins	2.4 kg
Triplets	1.82 kg
Quadruplets	1.4 kg

　Birth weight is influenced by:
• Social class
• Prematurity.

FETAL CIRCULATION

Blood from the aorta is sent in two directions:
1. A minor part goes to the lower limbs and then to the inferior
   vena cava (IVC).
2. A major part goes to the umbilical arteries and to the fetal side
   of the placenta.
   In the placenta blood travels to the umbilical veins:
• Through the liver to the IVC.

- Bypasses the liver (ductus venosus) to the IVC.
  The IVC contains blood from:
- Lower limbs
- Placenta, liver and alimentary tract.
  Blood from IVC diverted into two streams by the crista divides:
- The larger stream goes to the foramen ovale and into the left atrium.
- The smaller stream goes to the right atrium.
  The left atrium (blood from IVC and lung) goes to the:
- Left ventricle
- Aorta
- Brain
- Coronary arteries
- Descending aorta.
  The right atrium receives blood from the:
- Head via superior vena cava – low O_2 content.
- Coronary sinus – low O_2 content.
- Right stream from IVC – high O_2 content.
  The flow is directed to the:
- Right atrium
- Right ventricle
- Lungs (via pulmonary artery)
- Aorta (90%) through the ductus arteriosus.

EVENTS AT BIRTH (BASED ON THE SHEEP)

The events at birth are poorly understood in the human. However, the changes that occur in sheep have been well documented. After occlusion of the umbilical cord:

- Fetal blood O_2 falls.
- There are signs of hypoxia.
- Breathing is initiated by stimuli from skin and chemoreceptors.
- Cessation of umbilical flow→increased peripheral resistance.
- Rise in arterial blood pressure.
- Arterial O_2 content falls.
- Expansion of lungs and breathing begins.
- Fall in vascular resistance in the pulmonary artery.
- Increased Po_2 causes vasodilatation.
    Increased blood flow
    Reversal of flow through the ductus arteriosus→closure.
- Pressure in left atrium rises→foramen ovale closes.
- Within 1 min of the start of breathing, all venous blood which reaches the right atrium passes to the lungs.
- Valve of foramen ovale fuses with atrial wall (in about 1 week).
- Ductus venosus collapses.

LUNG LIQUID

Fetal lung secretes fluid to keep potential air spaces expanded. This is produced by active chloride transport. The rate of fluid formation is approximately 3–4 ml/kg/h. Once formed the fluid passes into amniotic fluid where it can be swallowed by the fetus. During labour there is a surge of the catecholamines which stimulates the absorption of fluid from the lungs, by activating a Na^+ channel and the sodium pump.

SURFACTANT (see also Ch. 3)

Surfactant is present in the lamellar bodies of type II alveolar cells from 24 weeks of gestation. It is secreted into lungs and amniotic fluid by week 30. Synthesis can be stimulated by:
- Glucocorticoids
- Thyroxine.
  Further secretions at birth are due to:
- Catecholamine surge
- Ventilation of the lungs with air.

LACTATION

STRUCTURE OF THE MAMMARY GLAND

The mammary gland is divided into lobules. Each lobule consists of alveoli which are the secretory elements. The alveoli connect to small ducts which lead to the lactiferous tubules, which in turn converge on the nipple. Beneath the areolae each tubule dilates to form an ampulla. The ampullae act as milk reservoirs. There are smooth muscle cells which surround the ducts and myoepithelial cells surround alveoli.

DEVELOPMENT OF MAMMARY GLANDS

The mammary glands are rudimentary until puberty. At puberty an enlargement occurs which is under hormonal control. Further enlargement occurs during pregnancy due to:
- Oestrogens and progesterone which promote development of ducts and alveoli.
- Growth hormone, chorionic somatomammotrophin.
- Prolactin required for full maturation and milk production.
- Adrenal corticoids and insulin are required but their function is permissive.

LACTATION

Lactation involves two processes:
1. Secretion of milk
2. Expulsion of milk.

Secretion of milk
Hormonally controlled.

Prolactin stimulates the alveoli to secrete milk.
High levels of oestrogen and progesterone inhibit the secretion of milk during pregnancy.
At parturition levels fall and allow prolactin to act.
Free flow of milk is not established until 3–4 days after delivery.
Suckling stimulates the continuous release of prolactin.
Prolactin levels are high during breast feeding, causing physiological amenorrhoea – ovulation is inhibited.
Regular removal of milk stimulates further flow.
Psychogenic factors can influence milk production.

Expulsion of milk
Suckling causes:
* Further milk production.
* Expulsion of milk from the ampullae.
    Suckling causes a reflex release of oxytocin from the posterior pituitary. This stimulates smooth muscle and myoepithelial cells. Milk is then expressed from the gland.

Colostrum
Colostrum is milk produced during the first 2–3 days. It is produced in small amounts only but is high in protein but low in fat and sugar. It also contains IgA, which protects the baby against infection.
    For the composition of milk see Chapter five. Milk of each species is specifically tailored to the nutritional and immunological requirements of the young of that species.

Benefits of breast feeding
As well as being more nutritionally important, breast milk adds protection for babies who in their first months of life are more vulnerable to infection.

Protection in advance of birth
Antibodies manufactured by the mother cross the placenta and circulate in the baby's blood for weeks or months.

Protection gained from breast milk
Antibodies (IgA) targeted at pathogens in the mother's (and baby's) immediate surroundings are formed by the mother and secreted in the milk. They bind to pathogens in the baby's gastrointestinal tract, preventing them from passing into the body. Agents which disrupt membranes of pathogens are:
* Lysozyme – kills bacteria by disrupting their wall.
* Fatty acids – disrupt membranes surrounding some viruses.
    Agents which deny pathogenic bacteria the substances required for their growth are:
* Vitamin B_{12}-binding protein.
* Lactoferrin – binds iron needed for growth on some organisms.

Agents which increase the antimicrobial activity of the immune system:
- Fibronectin – increases antimicrobial activity of macrophages.
- γ-Interferon – increases antimicrobial activity of immune cells.
  Agents which prevent the attachment to mucosal surfaces:
- Mucins – adhere to bacteria and viruses.
- Oligosaccharides – bind to microorganisms.
  Agents which stimulate the baby's gastrointestinal tract to mature more quickly:
- Cortisol
- Epidermal growth factor
- Nerve growth factor
- Insulin-like growth factor
- Somatomedin C.
  Agents which crowd out pathogenic bacteria from the gut:
- Bifidus factor – promotes the growth of *Lactobacillus bifidus*, a harmless bacterium which helps to crowd out pathogen.
- Cells secreted in the milk, particularly the colostrum
  B-lymphocytes – produce antibodies against specific organisms.
  Macrophages:
    Destroy pathogenic bacteria in the baby's gut
    Produce lysozyme
    Activate components of the immune system.
  Neutrophils – act as phagocytes in the gut.
  T-lymphocytes:
    Kill infected cells
    Mobilise components of the immune system.

In comparing bottle-fed and breast-fed babies, there appear to be as yet undiscovered factors which put babies at lower risk from infection when they are breast fed.

9. The kidney, body composition and acid–base balance

The kidneys have several functions:
1. The regulation of the volume and composition of the body fluids.
2. The production of humoral agents:
   a. Erythropoietin
   b. Metabolites of vitamin D
   c. Renin
   d. Prostaglandins.

THE NEPHRON

The fundamental unit of the kidney is the nephron, which is composed of a glomerulus and tubular system. In the glomerulus there is a capillary bed invaginated into Bowman's capsule. This is where blood is filtered. From Bowman's capsule there is a tubular system of epithelial cells which modify the filtered plasma (Fig. 9.1). These tubules are:
- Proximal convoluted tubule
- Loop of Henle
- Distal convoluted tubule
- Collecting tubules.

THE STRUCTURE AND FUNCTION OF THE GLOMERULUS

An afferent arteriole enters the glomerulus and divides into several loops (four to eight). The wall of each loop consists of a fenestrated endothelium, lying on a basement membrane with an epithelial layer next to Bowman's space. The endothelial layer has spaces (fenestrae) and the basement membrane is a complex of a loose fibrillar mesh of glycoproteins with fixed negative charges. Foot processes of the podocytes lining Bowman's space sit on the basement membrane. Between the capillary loops, forming a scaffolding, is the *mesangial* system, which may have a role to play in adjusting capillary blood flow. Slit pores between the foot processes provide a pathway from plasma to Bowman's space.

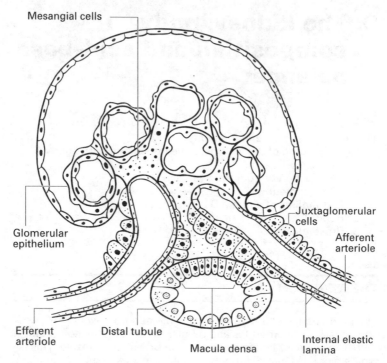

Fig. 9.1
Structure of the glomerulus.

VOLUME AND PROPERTIES OF URINE

The volume of urine excreted in 24 h depends upon:
• Fluid intake.
• Amount of solute cleared.
• The capacity of the kidney to concentrate or dilute the urine; in a normal adult about 800–2500 ml is excreted/24 h.

Only a small fraction of material filtered at the glomerulus appears in the urine. Typical figures for an adult man are:

Compound	Amount filtered (mmol/24 h)	Amount in urine (mmol/24 h)
Sodium	26 000	150
Chloride	18 000	200
Bicarbonate	4500	<2
Potassium	900	150
Urea	900	440
Glucose	900	<5

Of the 180 L of water filtered, only about 1.5 L is lost in the urine.

Colour
Yellow due to the pigment urochrome.

Protein
0–90 mg/day.
Transient proteinuria may occur after exercise.

Cells
Leucocytes and non-squamous epithelial cells (20 000–200 000/h).
Smaller numbers of red cells.
Some hyaline casts (precipitation of mucoprotein secreted by tubules).

GLOMERULAR FUNCTION

The first step in the formation of urine is ultrafiltration. The evidence for this comes from micropuncture studies sampling the fluid in Bowman's capsule. Analysis shows that this is virtually protein free. Ultrastructural examination of the glomerular membrane shows pores between the cells where only basement membrane is found. This membrane represents a poor diffusion barrier but the pore diameter limits the size of molecules that can pass. Typically, solutes below M_r 10 000 are freely filtered but as molecules increase in size and charge filtration is more difficult. Filtration stops for negatively charged albumin molecules of 70 000 and neutral solutes of 100 000 M_r.

The glomerular filtration rate (GFR)
The glomerular filtration rate is determined by physical forces. The arterioles in the kidney are connected directly to the aorta by short wide vessels and therefore can receive the maximum pressure possible from the heart. Arterioles before and after the capillary bed in Bowman's capsule (afferent and efferent arterioles) are capable of adjusting glomerular pressures and hence flow. Constriction of the precapillary arteriole decreases capillary pressure and therefore filtration. Constriction of the postcapillary arterioles increases capillary pressure and augments filtration. The forces involved are:

Driving force
Glomerular capillary hydrostatic pressure (P_g).

Opposing forces
Hydrostatic pressure in Bowman's capsule (P_{bs}).
Colloid osmotic pressure (*COP*) of the plasma proteins.
Permeability of the filtration membrane – including area (*K*).

Effective filtration pressure $= K (P_g - [P_{bs} + COP])$
125 ml are filtered by each kidney per minute.
All glomeruli are functional at all times.

Total capillary area = 1.5 m^2.
Number of glomeruli = 3×10^6.
Each glomerulus filters about 20–30 nl/min.

Filtration pressure
Putting some mean figures into the above equation:
P_b = 60 mmHg; P_{bs} = 20 mmHg; *COP* = 25 mmHg, perhaps a mean of 30 mmHg for the complete capillary, then:
Filtration pressure = 60 –(30 + 20) = 10 mmHg
This small driving force is made more effective by:
• A component from diffusion.
• Autoregulation of the renal blood flow; filtration remains constant over a range 80–60 mmHg.

Measurement of the GFR
This is an important measurement both for the physiologist and for the clinician assessing renal function. It can be assessed by measuring the clearance of a substance from the blood into the urine, which has the following properties:
• Biologically inert – not metabolised by the cells.
• Does not attach to a plasma protein.
• Freely filterable at the glomerulus.
• Neither secreted nor re-absorbed by the kidney tubules.
Substances used:
• Inulin – a polyfructose of molecular weight 5500.
• Cyanocobalamin (^{57}Co).
• EDTA (^{51}Cr).
• Creatinine – this is used for routine clinical purposes and as it is produced endogenously need not be infused intravenously.
To measure GFR it is essential to have a constant blood concentration of the substance used and be able to obtain accurate timed urine samples.
$$GFR = UV/P \text{ ml/min}$$
where:
P = plasma concentration in mg/ml
V = volume of urine produced ml/min
U = urine concentration mg/ml.
GFR is related to body surface area. The GFR for a normal male adult = 125 ± 15 ml/min/1.73 m^2. For females it is about 10% less.

Factors affecting the GFR
GFR is affected by:
• Size of an individual.
• Higher in the secretory phase of the menstrual cycle.
• Pregnancy – increases some 50–100%.
• Not constant but can vary from minute to minute.
• Renal plasma flow – an increase elevates the GFR proportionately.
• Constriction of afferent and efferent arterioles – renal plasma flow

and filtration pressures are affected and it may not be possible to predict the precise outcome.
- See effects of stimulating atrial receptors of the heart (pp 66, 296)
- Stimulation of the renal sympathetic nerves:
    Afferent arteriole constricts
    Decrease in filtration pressure
    Reduced GFR.

CONCEPT OF CLEARANCE

The clearance (C) of a substance (UV/P) is a measure of the minimal volume of plasma required to supply the amount of substance excreted in the urine in a given time:

C_{inulin} = GFR
C_{urea} < GFR – tubular re-absorption
C_{PAH} > GFR – tubular secretion.

MEASUREMENT OF RENAL BLOOD FLOW

If a substance is totally removed from the plasma on passing through the kidneys, so that none reaches the renal vein, then the clearance of that substance is equal to the renal plasma flow. Effective renal plasma flow (ERPF) is determined by measuring the clearance of para-aminohippuric acid (PAH), which is almost entirely removed from blood during one passage through the kidneys.

Method
A large dose of PAH is injected intravenously to give the required plasma concentration:
This concentration is maintained constant by the slow intravenous infusion of PAH in physiological salt solution.
Sufficient fluids are given by mouth to give a reasonable urine flow.
Accurately timed samples of urine are taken.
90% of the PAH is secreted in one passage through the kidneys.
PAH estimates the renal plasma flow.

$$ERPF = U_{PAH} \ V/P_{PAH}$$

In the young male the ERPF = 600 ml/min/1.73 m^2.
90% total flow – 10% passes through non-renal tissue.

$$\text{Renal blood flow} = \frac{\text{RPF (1)}}{(1 - \text{haematocrit})} \cong 1200 \text{ ml/min}$$

RBF = 20–25% of the cardiac output.
Filtration fraction = inulin clearance/PAH clearance.

DISTRIBUTION OF BLOOD FLOW IN THE KIDNEY

The distribution of blood in the different regions of the kidney is:

- Cortical peritubular capillaries 93%.
- Outer medullary capillaries 6%.
- Capillaries of papilla 1%.

AUTOREGULATION OF THE RENAL CIRCULATION

GFR and RBF change little with blood pressure over the range 80–160 mmHg. This represents an autoregulation of renal blood flow. Autoregulation is a property of the renal vasculature, since it is also functional in the denervated kidney. There are two theories to account for this mechanism:
1. May be myogenic, i.e. arteries contract when there is an increase in transmural pressure.
2. Tubuloglomerular feedback through the juxtaglomerular apparatus.

FUNCTION OF MESANGIAL CELLS

These cells are phagocytic. Their function is to engulf protein and immune complexes which escape from capillaries. Mesangial cells contain contractile elements which can affect glomerular capillaries by altering the area available for filtration and thereby alter the GFR. Vasopressin causes constriction of the mesangial cell and affects the amount of fluid filtered.

TUBULAR FUNCTION

The tubules in the kidney are different in structure and perform different functions.

Proximal tubule
Leaky epithelium.
Re-absorption of 60% of water and solute in isotonic proportions.
Glucose and amino acids are entirely re-absorbed.
The small amount of protein filtered is re-absorbed.
Hydrogen ions and ammonia secreted into lumen.

Loop of Henle
Re-absorbs salt and water.
Sets up interstitial concentration gradient.
Counter-current multiplier.
Fluid at tip of loop – hypertonic.
Fluid leaving loop – hypotonic.
Vasa recta – counter-current exchangers.

Distal tubule
Relatively tight epithelium.

Control mechanisms sited here – as the distal tubular system has a dominant effect on the composition of the urine.
Small amounts of solute absorbed against an electrochemical gradient.
Potassium, hydrogen ion and ammonia secreted.
Hypotonic at the beginning, isotonic at the end.

Collecting ducts
Many of the properties of the distal tubules.
Sodium re-absorption continues.
Collecting ducts pass through a hyperosmotic interstitium:
> Under a water load the tubules become impermeable to water→diuresis
> Under dehydration the tubules are permeable to water through ADH→concentrated urine.

SODIUM ABSORPTION

Sodium is the major ionic constituent of plasma. Consequently, the ability of the kidney to absorb sodium is important. Ninety-nine per cent of the filtered sodium load is re-absorbed. The different sections of the kidney carry this out to different extents:
- 60% absorbed in proximal tubule.
- 10% of filtered load absorbed in distal tubule.
- 2% of distal absorbate under control of aldosterone.

FACTORS CONTROLLING SODIUM SECRETION – RE-ABSORPTION

Glomerular filtration rate (GFR)
An increase in load to tubule increases the rate of absorption (glomerulotubular balance).

Aldosterone
The secretion rate is regulated by:
- Sodium balance (through osmolality effect):
  > High sodium intake→low aldosterone secretion
  > Low sodium intake→high aldosterone secretion.
- Hypovolaemia (e.g. haemorrhage) increases aldosterone secretion→Na retention.
- Plasma [K^+].
- ACTH.

Naturetic factor
This diminishes re-absorption in the distal tubule.

Physical factors (Starling forces acting on peritubular capillaries)
Hydrostatic pressure.
Colloid osmotic pressure.

Intrarenal distribution of glomerular filtration

Superficial nephrons 'salt losing'.
Juxtamedullary nephrons 'salt retaining'.

NEURAL INPUT TO THE KIDNEY

The kidney is supplied by nerves from the sympathetic nervous system. The function of these nerves is to influence the distribution of blood flow between the superficial nephrons and the juxtamedullary nephrons. There is an extensive adrenergic innervation of:

- Afferent and efferent glomerular arterioles.
- Proximal and distal tubules.
- Ascending limb of loop of Henle.
- The juxtaglomerular apparatus.
- Renal sympathetic nerves regulate urinary sodium by:
    Changing renal haemodynamics (e.g. GFR)
    Changing renin release from juxtaglomerular apparatus
    Changing tubular sodium re-absorption by:
        Direct action on renal tubules
        Indirectly through angiotensin II (as a result of sympathetic action).

Activation of renal sympathetic nerves can be brought about by *cardiorenal reflexes*. This involves receptors in atria which respond to stretch. Afferent information is then carried by the vagus nerve to the cardiovascular centres in the medulla. Efferent sympathetic nerves carry output information to the kidneys. This system senses changes in blood volume as a change in atrial pressure. Altered activity in the renal nerve changes the distribution of renal blood flow, capillary pressures in Bowman's capsule and thus re-absorption of sodium.

The renin–angiotensin–aldosterone–kinin system, showing the pathways of action and the points at which angiotensin-converting enzyme inhibitors and angiotensin II receptor antagonists act, is illustrated below.

HORMONES WHICH AFFECT SODIUM TRANSPORT

Prostaglandins
Bradykinin
Glucocorticoids
Thyroxine
Insulin
Growth hormone
Prolactin.

POTASSIUM SECRETION

Potassium is filtered and re-absorbed but can also be secreted by

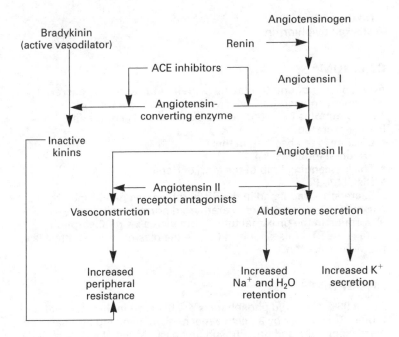

the tubules into the urine. Potassium contributes approximately 7% of the filtered load and is re-absorbed in the proximal tubule. The passive secretion occurs in the distal tubules and is determined by:

- Exchange for Na-aldosterone.
- Intracellular potassium concentration.
- K^+ and H^+ which compete in the excretory process.
- Potassium deficiency→low intracellular concentration→ H^+ secretion→alkalosis.

GLUCOSE SECRETION

Glucose is re-absorbed in the proximal tubule.
Renal threshold for glucose: at concentrations > 10 mmol/L appears in the urine.
Tubular maximum (TM) ≅ 2.0 mmol/min (maximum amount of glucose absorbed per minute). TM is influenced by:
  GFR
  Rate of proximal water and salt re-absorption.
Glucose transport:

  An active mechanism
  Carrier mediated
  Capacity limited

Na dependent
Blocked by phlorizin.

CALCIUM

About 50% is bound to plasma protein and cannot be excreted. It is mostly ionised Ca^{2+} which is excreted and a total of 1.5–7.5 mmol is excreted per day. Less than 1% of the filtered load is excreted.

Calcium is re-absorbed in the:
- Proximal tubule (60%).
- Thick ascending limb of the loop of Henle.
- Distal tubule.

Excretion is regulated in the distal tubule and determined by the plasma Ca^{2+} concentration. Parathyroid hormone inhibits re-absorption in the proximal tubule, but increases re-absorption distally. As PTH raises ionised Ca^{2+} in the plasma, the net effect is to promote excretion.

PHOSPHATE

About 95% of filtered phosphate is re-absorbed in the proximal tubule. This occurs by an active mechanism involving luminal cotransport with sodium. Phosphate has a TM but at a level slightly less than plasma concentration, therefore there is always some phosphate in the urine where it acts as a buffer. The amount re-absorbed depends upon:
- Phosphate concentration in plasma – low concentrations favour absorption.
- Parathyroid hormone – excess activity (hyperparathyroidism)→inhibition of tubular re-absorption →urinary loss.

BICARBONATE

About 4500 mmol of bicarbonate is filtered per day. The secretion of H^+ into the lumen throughout the nephron converts HCO_3^- to CO_2 and H_2O. However, for every H^+ secreted one HCO_3^- is synthesised by the tubular cells and returned to the peritubular capillaries (see p. 316). The balance of the process is re-absorption of all the bicarbonate (on a mixed diet):
- 70–85% in the proximal tubule.
- 10–20% in the thick ascending limb of the loop of Henle.
- 3–5 in the distal tubule.
- 1–2% in the collecting ducts.

During alkalaemia, plasma bicarbonate concentration is elevated and the re-absorptive process is swamped. As a result bicarbonate appears in the urine.

TUBULAR TRANSPORT MECHANISMS

Cations

Sodium
Sodium can cross the luminal membrane:
1. Passively down its electrochemical gradient.
2. By cotransport coupled to glucose.
3. By cotransport coupled to amino acids.
4. In exchange for H^+.
   Sodium leaves the cell across the basolateral membrane by active transport (against an electrochemical gradient) in exchange for K^+. This process uses the Na^+–K^+ ATPase.

Anions

Bicarbonate
Crosses the tubular membranes as CO_2.
Enters the peritubular capillaries as bicarbonate.

Chloride
Chloride moves passively through intercellular channels in exchange for formate. The latter, when it enters the lumen, reacts with H^+ to form the weak acid, formic acid, which because it is lipid soluble (in contrast to its anion) diffuses back into the cell.

The loop of Henle
The loop of Henle is composed of three parts:
1. Thin descending limb.
2. Thin ascending limb.
3. Thick ascending limb.

The counter-current system
By the time fluid has reached the loop of Henle it has been reduced from 180 L/day to about 60 L/day, containing 9000 mmol Na^+. A region of gradually increasing hyperosmolality is set up from the junction of the cortex and medulla to the tips of the renal papillae (285–1400 mosmol/kg). Only species with loops of Henle can produce a concentrated urine.
   The gradient is established by Na^+ and Cl^-, moving out of the ascending limb into the interstitium. This occurs by active cotransport from the thick ascending limb and passive cotransport from the thin ascending limb. These parts of the loop are impermeable to water so that the contents of the lumen become hypotonic as they approach the cortex. The thin descending limb is permeable to water, but less so to Na^+ and Cl^- and urea, so that water leaves the tubule and its contents become more concentrated.

The role of the vasa recta
These are capillaries which send loops down into the medulla and

whose descending and ascending limbs are in close contact with each other. Re-absorbed solute in ascending limbs coming from hyperosmotic medulla will be greater than that in the descending limbs and solute will diffuse into the descending limb, trapping solute. By the same reasoning, water in the more dilute plasma of the descending limb will pass into the ascending limb, thus acting as *counter-current exchangers*.

The concentrating and diluting kidney
In the basal state GFR = 125 ml/min. This results in a urine flow of approximately 1 ml/min. Thus 90% of the fluid filtered is absorbed.

The response to a water load
If a large volume of water is drunk rapidly ($\cong$ 1 L) there is an increased urine flow (*diuresis*), which begins after about 20 min. There is a low circulating level of ADH. Urine flow is maximal after about 1 h, at which time the osmolality of urine falls to about 85 mosmol/kg. The water load is eliminated in 2–4 h. During this time the excretion of solute and GFR are unchanged.

Response to deprivation of water for 12 h
During water deprivation urine output falls below 1 ml/min. The osmolality of the urine rises and may reach 800 mosmol/kg. Despite this, the osmolality of body fluids is kept approximately constant at 285 mosmol/kg.

The role of antidiuretic hormone (ADH)
Vassopressin is released from posterior pituitary in response to:
• Dehydration
• Pain
• Diminished cardiac output
• Cigarette smoking.

Osmotic control of ADH release
An increase in plasma osmolality increases ADH release. The threshold to initiate ADH release is a plasma osmolality of 280 mosmol/kg. This affects osmoreceptors in the hypothalamus, which trigger release of ADH from the posterior pituitary.

Mode of action of ADH
ADH binds to the basolateral membranes of the distal tubule and the cortical and medullary portions of the collecting ducts. There it activates adenylate cyclase and intracellular cAMP rises. cAMP acts through protein kinases which lead to the fusion of cytoplasmic vesicles with the luminal membrane. These vesicles contain water channels and this new membrane increases the water permeability and water moves down its osmotic gradient from tubular lumen to interstitial fluid (antidiuresis). The permeability of inner medullary collecting ducts to urea is increased. Urea is recycled between

medullary interstitium and tubular fluid, allowing the medullary hyperosmolality to be preserved. Urine formation can be reduced to about 0.5 L/24 h (0.3 ml/min) and osmolality may reach 1200 mosmol/kg. In the absence of ADH, in the condition known as diabetes insipidus, urine flows of 20–30 L/24 h can occur ($\cong$ 20 ml/min at an osmolality of about 50 mosmol/kg).

TESTS OF RENAL FUNCTION

Tests of renal function are based on *clearance tests*. These tests determine the ability of the kidney to concentrate specific substances in the urine. For example:
- Creatinine
- Inulin
- Urea.

  Typically the following procedure is used:
- Bladder emptied.
- No food or fluid for 8 h.
- Bladder emptied after 12 h.
- Osmolality of specific gravity determined.
- Vasopressin tannate (ADH) injected intramuscularly.
- Fluid withheld a further 6 h.
- Osmolality of urine formed tested – should be about 900 mosmol/kg.
- Test for protein in the urine.

REDUCTION IN MAXIMUM CONCENTRATING ABILITY

In certain clinical conditions the kidney has a reduced capability to concentrate urine. This can be due to:

- ADH deficiency – nephrogenic diabetes insipidus.
- Structural damage to renal tubules:
    Repeated infections
    Phenacetin – drug-induced damage.
    Hypokalaemia
    Hypercalcaemia.
- True diabetes insipidus responds to injected ADH.
- Nephrogenic diabetes insipidus does not respond to ADH.

THE RENIN–ANGIOTENSIN–ALDOSTERONE SYSTEM

Renin is a proteolytic enzyme with a molecular weight of 40 000. It originates at the juxtaglomerular apparatus where it is stored within cells. Renin hydrolyses angiotensinogen (an α_2-globulin synthesised by the liver) to form angiotensin I (a decapeptide). Angiotensin I is converted into angiotensin II (an octapeptide) by converting enzyme. The converting enzyme is found in the kidney and lung. Angiotensin II has a number of actions:

Intrarenal actions
Stimulates Na re-absorption by the nephrons (proximal tubule).
Constricts glomerular arterioles→reduces GFR.

Extrarenal actions
Stimulates the release of aldosterone from the zona glomerulosa of the adrenal cortex.
Constricts blood vessels→increased peripheral resistance→increase in blood pressure.
Stimulates the release of ADH.
Stimulates thirst.
Probably involved in the long-term regulation of the blood pressure in patients with hypertension; an inhibitor of converting enzyme reduces blood pressure just as in the case of blocking the binding of angiotensin II to its receptor on the zona glomerulosa cells.

Physiological role of renin
The stimuli for renin release include:
• Decrease in blood pressure acting on afferent arterioles (reduction in perfusion pressure).
• Adrenergic nerve stimulation acting through β-adrenoreceptors.
• Circulating catecholamines.
• A decrease in the rate of perfusion of tubular fluid passed through the macula densa.

Hypertension and the renin–angiotensin system
The kidney can be associated with several forms of hypertension:
• Hypertension can result as a consequence of an excess production of aldosterone by cells in the zona glomerulosa.
• In cases of renal artery stenosis there can also be an excess renin production producing hypertension.
• Malignant hypertension (excess renin and aldosterone) may be secondary and not the cause.
• Essential hypertension – no abnormality so far discovered.
   Hypertension can be treated. One way is to use drugs which interfere with the renin–angiotensin pathway.

ATRIAL NATURETIC PEPTIDES (ANP)

Atrial naturetic peptide (ANP) has 28 amino acids and is stored in granules in the cardiac myocytes of the right atrium. The granules are released in response to an increase in atrial pressure. ANP has the following actions:
• Infusions cause a fall in blood pressure→arteriolar vasodilatation.
• Venodilatation.
• Decreases the sensitivity to vasoconstrictor substances – noradrenaline and angiotensin II.
• Decreases the activity of the renin–angiotensin–aldosterone system.

- Directs the inhibiting effect on the zona glomerulosa of the adrenal gland.
- Relaxes arterial walls.
  ANP also has effects on the kidney:
- Increases glomerular filtration rate.
- Inhibits tubular Na^+ absorption.
- Increases urine flow→naturesis.
- Decreases release of ADH→diuresis.

THE KIDNEY AS AN ENDOCRINE ORGAN

The kidney produces:
- Renin.
- Erythropoietin (stimulates red cell production) released by:

    Hypoxia
    Anaemia
    Administration of androgens.
- 1,25-Dihydroxycalciferol (active vitamin D).
- Vasodilator and naturetic factors:

    Prostaglandins PGA_2, PGE_2
    Kallikrein
    Kininogen$\xrightarrow{\text{Kallikrein}}$bradykinin.

RENAL FUNCTION IN THE NEWBORN

The neonatal kidney is less functional than the adult kidney. GFR is approximately 30 ml/min/m^2, resulting in 20 ml/day filtered in the newborn. As a result of this low GFR, a water load is excreted slowly. The superficial nephrons are underdeveloped, but their function increases as:

- Arterial blood pressure rises.
- A fall in intrarenal vascular resistance occurs.
- Glomerular permeability increases.

GFR and blood flow are greater in juxtamedullary nephrons. The juxtamedullary nephrons retain salt and the neonate is unable to excrete a saline load. As a result, there is a high capacity to retain both Na^+ and water. Tubular function is less than that of the adult. The kidney is fully developed and functional by the end of the first year.
  In the neonate, the peritubular physical factors are altered:

- Oncotic and hydrostatic pressures, which are reduced.
- Osmotic gradients, which are reduced by high blood flow through the vasa recta.
  As a result of this, the concentrating ability of the nephron is

reduced to approximately 700 mosmol/kg, about maximum. A lack of ADH is not involved.

RENAL FUNCTION IN THE ELDERLY

In the elderly renal function diminishes. This begins after the 4th or 5th decade. At this time the GFR falls to 40–70%. The renal blood flow falls and the concentrating capacity of the kidney is reduced. Obstructive vascular disease becomes important. Furthermore, the number of glomeruli in each kidney falls from 2.1×10^6 to 1.0×10^6.

MICTURITION

Micturition is the process of emptying the bladder. Urine collects in minor calyces from the collecting tubules and coordinated contractions in the circular muscle layer propel urine down ureters. Contractions travel 2–3 cm/s with one contraction about every 10 s. It has been suggested that there may be a pacemaker in the calyces. Urine enters the bladder in spurts. The smooth muscle has no nerve supply but has α-adrenergic receptors. There is a valvular opening into the bladder.

THE URINARY BLADDER

The sphincters controlling urine outflow of urine are tonically contracted when the bladder is empty. The detrusor muscle and internal sphincter (neck of bladder) are innervated by:

1. The sympathetics from the lumbar segments of the spinal cord – inhibitory.
2. The parasympathetics from segments S2 to S4 – excitatory.
   The external sphincter is skeletal muscle and, as such, has somatic innervation.

During filling, the detrusor muscle relaxes and the pressure rise is at first very small. Awareness of bladder filling occurs when the volume of urine reaches about 150 ml. The bladder can accommodate up to 300–400 ml with little increase in pressure and tension in the bladder wall. When about 600 ml has collected, a feeling of urgency arises and contractions are initiated in the detrusor muscle. The pressure begins to rise to about 100 cmH$_2$O. Voluntary restraint can occur and pressure falls temporarily. However, reflex micturition is inevitable.

REFLEX NATURE OF MICTURITION

As bladder fills afferent impulses pass via the pelvic splanchnics to a reflex centre in the spinal cord. Impulses from this centre pass via the pelvic splanchnic nerves to cause:

- Detrusor muscle contraction.
- Internal sphincter relaxation.

Efferent impulses in the pudendal nerve cause opening of the external sphincter and the passage of urine. Urine flow through the external urethra causes afferent fibres in the pudendal nerve to discharge, which reinforces parasympathetic excitation through the supraspinal mechanism. The increase in bladder pressure produced by the autonomic discharge causes a self-reinforcing cycle by:

- Further afferent discharge from tension receptors in the bladder wall.
- Reflexly exciting efferent discharge in the parasympathetic nerve fibres.

ROLE OF HIGHER CENTRES

Higher centres in the brain also influence micturition. Inputs from the cerebral cortex, basal ganglia and reticular formation:

- Inhibit the sacral reflex centre during bladder filling.
- Inhibit the sacral centre when voluntary restraint is required.
- Facilitate the sacral centre during micturition.

Afferent fibres conveying information of awareness of bladder filling travel in:

- Pelvic splanchnics.
- Sympathetic (hypogastric) nerves.

VOLUNTARY MICTURITION

During voluntary micturition the inhibition of the sacral reflex centre is removed. This causes relaxation of perineal muscles. The sphincter muscles do not relax until the detrusor contracts. The process is augmented by contraction of the abdominal muscles, resulting in an elevation of intra-abdominal pressure.

VOLUNTARY RESTRAINT

It is possible to prevent micturition. This involves contraction of perineal muscles and continued inhibition of the sacral reflex centre. The external sphincter is kept closed.

DISORDERS OF MICTURITION

Disorders of micturition occur with transverse lesions of the spinal cord, e.g. paraplegia. In such conditions reflex micturition is at first abolished and the detrusor relaxed. The bladder becomes overdistended with urine, overflow occurs and the subject is incontinent. After some time the patient micturates automatically and has a smaller bladder capacity. A small increase in urine volume gives rise to a large increment in pressure. The detrusor contracts about once per hour.

THE BODY FLUIDS

BODY FAT

As the body fat has considerable influence on body water, it is convenient to consider it here. The fat content of the body can be assessed in qualitative and quantitative ways.

ASSESSMENT BY EYE

Is the person average, thin, plump or fat?

BODY MASS INDEX (QUETELET INDEX)

Body mass index (BMI) = Weight (kg)/Height (m)^2
There are four grades of BMI:

Grade 0   < 25
Grade 1   25–29.9
Grade 2   30–40
Grade 3   > 40
A BMI grade greater than 30 is regarded as abnormal.

SKIN-FOLD THICKNESS

A simple way of assessing body fat can be made by measuring the thickness of the folds of skin at agreed sites with callipers:
1. Biceps – mid-point of muscle with arm hanging vertically.
2. Triceps – at a point equidistant from the tip of the acromion and the olecranon.
3. Subscapular – just below the tip of the inferior angle.
4. Suprailiac – over the crest in the midaxillary line.

BODY DENSITY

For men
Density = 1.1610 – 0.0632 log sum of skinfold thickness at four sites
For women
Density = 1.1581 – 0.0720 log sum of skinfold thickness at four sites
Insert densities into equation below.
A more fundamental method is that of measuring body density. This is done by first weighing the subject in air and then under

water. A snorkel is used so that the residual volume can be measured using a nitrogen dilution method.

$$\text{Density} = \text{Mass/Volume}$$

The volume is obtained from the weight of water displaced corrected for the temperature of the water. In order to calculate the fat composition, let X be the percentage of fat in the body, then:

$$1/\text{Density of body} = (1{-}X)/\text{Density of lean body mass (LBM)} + 1/\text{Density of fat}$$

The density of human fat is 900 kg/m^3 and that of the fat-free body is 1100 kg/m^3.

$$1/\text{Density of body} = 1{-}X/1100 + X/900$$

$$\therefore X = \frac{4950}{D} - 4.5$$

or FFM (fat-free mass) = X × Body weight

Note that in measuring density a correction has to be made for the upthrust of air in the lungs, i.e. the residual volume (RV).

The body density (D) (in kg/m^3) is given by:

$$D = \frac{\text{Mass in air}}{[(\text{Mass in air} - \text{Mass in water})/\text{Density of water}] - RV}$$

RV is the residual volume and is measured by a nitrogen dilution method (see Ch. 3) at the same time as the weight is taken. RV is litres at body pressure, ambient temperature, saturated with water vapour (BTPS), and is given by:

$$RV = 3000 \times \frac{(N - 0.5)}{(80 - N)} \times \frac{Bp}{(Bp - 47)} \times \frac{310}{(273 + t)} DS$$

where DS = dead space of snorkel and tap, Bp = barometric pressure, t = temperature. The nitrogen content of alveolar air is assumed to be 80%.

ELECTRICAL IMPEDANCE

As fat has a high specific resistance, the fat content of the body can be obtained by measuring the resistance of sections of the body to small high frequency currents. This method has to be calibrated against the body density method.

Body fat, LBM, water expressed as a % of body weight for a thin and fat man and a thin and fat woman

	% fat	% LBM	% TBW
Thin man	4	26	70
Fat man	32	18	50
Thin woman	18	22	60
Fat woman	42	16	42

(Figures reproduced from Smith & Brain Fluids and Electrolytes Pub. Churchill Livingstone 1980)
The large variations in TBW (total body water) are due mainly to the proportion of fat in the body.

TOTAL BODY WATER (TBW)

The amount of water in the body depends upon:
• Weight of individual
• Age
• Sex
• Fat content.

Weight
The total body water is about 60% (42 L) of the body weight in an average man, but can vary from 45 to 75%. It is also relatively constant at 73% of the LBM, i.e. fat-free tissue.

Age and sex
The content of water falls with age and women of the same age as men have a lower water content after puberty.

Fat content
As fat is relatively free of water, the water content of the body varies with the degree of fatness.

Total body water as a percentage of body weight

	Infant	Male	Female
Thin	80	65	55
Average	70	60	50
Fat	65	55	45

COMPOSITION OF BODY FLUIDS

Body fluids are compartmentalised. The major divisions are the extracellular and intracellular compartments.

Extracellular fluid (ECF)
The ECF is the water content outside the cells. The ECF contains one-third of the TBW (about 20% of the body weight or 14 L). It has two major divisions:

1. The *interstitial fluid* (ISF) bathing the cells (75% of the ECF) or 15% of body weight ($\equiv$ 10 L).
2. The *plasma* (25% of the ECF) confined to the blood vessels or 5% of body weight ($\equiv$ 3 L).

Intracellular fluid (ICF)
The ICF is the sum of the water content of each cell or about 40% of the body weight ($\equiv$ 28 L).

The figures above are for an average man of 70 kg. Slightly different figures are given in various books. The reason for this is that different indicators have been used in the measurements and for most clinical purposes the water in dense connective tissue and bone may have been discounted because these compartments only exchange with the main ICF very slowly.

Measurement of the volume of the body fluid compartments

The volumes of the ECF and the ICF compartments are measured by indicator dilution techniques. The principle of this approach is that the volume of distribution of a water-soluble substance when added to a compartment is determined by the following equation:

$$Q = V \times C$$

where:

Q = the quantity of substance added to a compartment
V = is the volume of distribution within that compartment
C = is the concentration in the compartment when the substance is evenly distributed.

An indicator must have certain properties:
1. When injected intravenously it must be confined to the compartment.
2. It must be non-toxic.
3. It is neither synthesised nor metabolised.
4. If excreted it must be at a constant rate.
5. It should be capable of accurate measurement.

Measurement of ECF space

Suitable markers are:
- Inulin
- Sucrose
- Thiosulphate
- Thiocyanate.

A dose of indicator, e.g. inulin, is injected intravenously, calculated to give an approximately suitable blood level. Blood samples are taken at intervals and the plasma concentrations measured. The result when plotted gives a double exponential curve. When the falling phase (due to loss in the urine) is plotted on a logarithmic scale, a straight line results. Extrapolation of the line to cut the axis at zero time gives the concentration of the marker in plasma which would have occurred if it had been evenly distributed instantaneously. This value is put into the equation:

$$\text{Volume of ECF space} = \frac{\text{Quantity injected}}{\text{Concentration at zero time}}$$

Measurement of total body water

Suitable indicators or markers must penetrate cell membranes such as:

- Heavy water

- Titrated water
- Antipyrine.

A known quantity (Q), made up in isotonic NaCl, is injected intravenously and samples taken until the concentration (or counts) in the blood are constant (at equilibrium), C. Urine is collected over the period and the amount of marker excreted measured (q).

$$\text{The volume of distribution (TBW)} = \frac{(Q - q)}{C}$$

Measurement of cell water (ICF)

There is no direct measurement of this ICF and so it is obtained by difference.

$$\text{ICF} = \text{TBW} - \text{ECF}$$

The following table gives typical values for the composition of ICF and ECF.

Composition of the body fluids

	Extracellular concentration fluid	(mM)	Intracellular concentration fluid	(mM)
Principal cation	Na^+	140	K^+	160
Other cations	K^+	4	Na^+	10
	Ca^{2+}	2.6	Ca^{2+}	Mostly bound
	Mg^{2+}	1.0	Mg^{2+}	13
Principal anion	Cl^-	100	$Pr^- + PO_4^{2-}$	187
Other anions	HCO_3^-	27	HCO_3^-	10
	$Pr^- + PO_4^-$	11	Cl^-	10

Measurement of sodium and potassium

Total amounts of Na^+ and K^+ are usually determined by isotopic dilution. $^{42}K^+$ occurs naturally and is a constant proportion of the total K^+ and is measured by total body scanning. $^{22}Na^+$ is injected intravenously as isotonic saline and its activity determined in a blood sample after equilibration. It is 17% lower than chemical analysis because only about one-third of sodium is immediately exchangeable. Bone Na^+ has a long turnover time.

Daily turnover of water and electrolyte

In normal circumstances there is always a balance between input and output of water. A healthy person in a temperate climate at rest can maintain water balance on about 650 ml/day.

Note that water loss by all routes is accompanied by sodium, except from the lung where it is a pure water loss.

Daily turnover of sodium and potassium

	Na (mmol/day)	K (mmol/day)
Intake		
Food	150	100
Output		
Urine	80–100	30–100
Sweat	4	5
Faeces	1.5	4

CONTROL OF EXTRACELLULAR VOLUME AND ELECTROLYTE CONCENTRATIONS

The volume and osmolality of the ECF are monitored and have separate but integrated roles to play in a homeostatic mechanism which keeps the volume of the ECF relatively constant. Water balance is regulated primarily by changes in the volume of water ingested and the amount of urine lost, which in their turn are controlled by thirst and the concentration of antidiuretic hormone (ADH) in the plasma (Fig. 9.2).

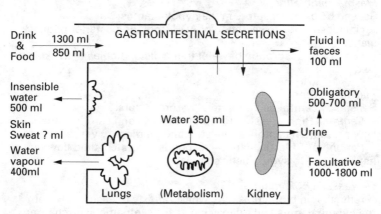

Fig. 9.2
Water balance.

CONTROL OF WATER INTAKE

The sensation of thirst is an important psychological factor in water intake. The sensations of thirst can be stimulated and inhibited.

Stimulation
An increase in the osmolality of the ECF.
A decrease in circulatory volume.
The renin–angiotensin system.
Dry mouth and pharynx.
Social habits.

Inhibition
Stomach distension.

Role of the kidney in changes in blood volume
The kidney plays a central role in the regulation of body water.

Increase in blood volume
Glomerular filtration rate (GFR) increases.
Tubular re-absorption decreases.
Fluid and salt loss occur.

Decrease in blood volume
GFR decreases.
Tubular re-absorption increases.
Fluid and salt retention.
   The factors and mechanisms which alter body water content
include:

Plasma osmolality
Sensed by osmoreceptors in the hypothalamus.
An increase leads to secretion of ADH from the posterior pituitary
gland.
Leads to a re-absorption of water in the distal tubules of the kidney
(see p. 300).

Plasma volume
Sensed by central and peripheral baroreceptors.
   Central   High-pressure receptors in aorta and carotid sinus.
Low-pressure receptors in the atria and pulmonary vessels.
   Peripheral   Macula densa – responds to stretch, fluid flow in
nephron and to sympathetic stimulation.

Decrease in ECF volume
A reduction in blood pressure, which reflexly releases ADH from
the hypothalamus and reflexly increases sympathetic discharge,
leads to vasoconstriction and a redistribution of glomerular
filtration from the superficial glomeruli (salt losing) to the
juxtamedullary glomeruli (salt retaining). There is also an increase
in proximal tubular re-absorption. Renin is released from the
macula densa, affecting the blood vessels and giving rise to a
reduction in perfusion pressure. Sympathetic nerve action further
leads to a reduction in GFR, thus reducing tubular flow past the
macula densa. There is a direct action on juxtamedullary apparatus
by the sympathetics.

The effects of renin secretion

Renin $\longrightarrow$ Angiotensin I $\longrightarrow$ Angiotensin II $\rightarrow$ **Thirst**
   Angiotensinogen   Converting enzyme   **Vasoconstriction**
                                                 $\downarrow$
                                          Aldosterone
                                                 $\downarrow$
                             Tubular re-absorption of Na^+
                             Secretion of K^+
                                                 $\downarrow$
                             Water and electrolytes conserved
                             by the kidney

Integrated control of aldosterone secretion

There is a further interaction between volume and osmotic control mechanisms at the level of the adrenal cortex (zona glomerulosa). A rise in plasma osmolality decreases the secretion of aldosterone and so reduces the re-absorption of sodium by the kidney, thus promoting sodium loss and correcting for the rise in osmolality.

The regulation of sodium

The sodium concentration in the plasma is usually kept within fairly close limits, yet there is no receptor which specifically monitors sodium. Sodium salts form the major ions in plasma and therefore the total osmolality of plasma, P_{osm}, is approximately equal to 2 × plasma sodium concentration. Because of this, the control of osmolality and volume of plasma must automatically control the sodium. In addition to these mechanisms, a third is also involved which is hormonal. Atrionaturetic peptide is released from atrial myocytes when right atrial pressure is raised. It acts on the distal tubular system of the kidney to inhibit the re-absorption of sodium and therefore causes a naturesis to reduce the sodium content of plasma.

DERANGEMENTS OF SALT AND WATER BALANCE

Saline depletion

Causes
Loss of salt and water in approximately physiological concentrations from:
• Alimentary tract.
• Kidneys.
• Skin – severe burns.
• Reduced intake – rare.

Consequences
Redistribution of fluid between cells and ECF.
Loss mainly by ECF (cells contain only little sodium).

Clinical signs
Diminished skin turgor
Dry tongue
Muscle cramps
Hypotension
Circulatory failure.

Saline overload

Causes
Overtreatment with intravenous saline.
Heart disease→failure to excrete daily salt load.
Renal disease→failure to excrete daily salt load.

Water depletion (inadequate intake)

Causes
Inadequate intake after losses with a low sodium content.
Sweating.
Pure water loss due to hyperventilation.
Loss of water in excess of sodium – diabetes insipidus (defect in ADH secretion).

Consequences
Water moves out of cells to dilute the ECF.
Cells tend to shrink – this affects intracellular metabolism, e.g. brain.
Concentrated urine, except in:
  Diabetes insipidus
  Nephrogenic diabetes insipidus.

Water excess (water intoxication)

Causes
Water intake exceeds the secretory capacity of the kidney, e.g. over 20 L/day in an otherwise healthy individual.
Excessive release of ADH occurring:
  After surgery
  From ectopic (tumour) sources.
Intravenous therapy with dextrose solutions.

Consequences
[Na^+] of plasma low.
Water diffuses rapidly into cells leading to cell swelling.
In the CNS this gives rise to:
  Nausea and vomiting
  Confusion
  Death.

Potassium deficiency (hypokalaemia)

Causes
Repeated vomiting.
Chronic diarrhoea (laxative abuse).
Urinary loss:
  Excess secretion of mineralocorticoids – hyperaldosteronism
  Diuretic drugs.
Low sodium diet:
  Due to re-absorption of sodium in exchange for potassium.

Consequences
Heart and skeletal muscle weakness, leading to paralysis if severe.

Potassium excess (hyperkalaemia)

Causes
Renal disease – potassium is retained.
Excessive oral or intravenous potassium.
Hypoaldosteronism.
Acidosis causes a loss of potassium from the cells.

Consequences
Interference with muscle action, particularly the heart, resulting in:
  Flaccid paralysis
  Cardiac arrhythmias.
Tall peaked T waves on the ECG.
Treatment very urgent.

ACID–BASE BALANCE

The regulation of the amount of acid and base is very important in the body, as they affect many enzymes and processes. Normally this is discussed in terms of the hydrogen ion concentration. The term pH is defined as the negative logarithm of the hydrogen ion concentration.

HYDROGEN ION CONCENTRATION

Hydrogen ions in arterial blood:
- Normal [H^+] in blood 36–44 nmol/L or pH = 7.44–7.36
- Extreme limits in disease 20–126 nmol/L or pH 7.7–6.9
- Wide variations can only be tolerated for a very short time.

PRODUCTION OF ACID

On a mixed diet acid is produced as a result of:

Oxidation of carbon compounds
CO_2 approximately 13 000 mmol/24 h.
Organic acids lactic and pyruvic acids.

Oxidation of sulphur and phosphate compounds – about 70 mmol/day
Methionine and cysteine→H_2SO_4.
Organic phosphates→H_3PO_4.

CONTROL OF [H^+]

The concentration of H^+ is controlled by buffering which is immediate, short term, and is the first line of defence.

DEFINITION OF A BUFFER

A buffer system minimises the change in $[H^+]$ when acid or a base is added to it:

$HA \Leftrightarrow H^+ + A^-$ (HA = undissociated buffer: A^- = anion base)

Addition of H^+ shifts the reaction to the left.

Anion accepts H^+ and undissociated buffer concentration [HA] increases and limits the increases in $[H^+]$ with only a small change in arterial pH.

Removal of H^+ or addition of base shifts the reaction to the right, undissociated buffer now dissociates and contributes H^+ which reduces the fall in $[H^+]$, i.e. only a very small rise in arterial pH occurs.

BUFFER SYSTEMS OF THE BODY

The body has several buffering systems:

$H_2CO_3 \Leftrightarrow H^+ + HCO_3^-$
*HHb $\Leftrightarrow H^+ + Hb^-$
HPr $\Leftrightarrow H^+ + Pr^-$ (Pr = protein)
$H_2PO_4^- \Leftrightarrow H^+ + HPO_4^{2-}$

Note: *HHb is most important in gaseous transport by the red cell.

ROLE OF THE LUNGS

The lungs provide a second line of defence which is rapid in action and operates to bring about a fine adjustment of $[H^+]$. This is achieved using CO_2 and HCO_3.

$$CO_2 + H_2O \Leftrightarrow H_2CO_3 \Leftrightarrow H^+ + HCO_3^-$$

The removal of the CO_2 by the lungs drives the reaction to the left and the carbonic acid formed by bicarbonate buffering is removed.

ROLE OF THE KIDNEY

This is the third line of defence against acid. It is slower to act and is a long-term compensatory mechanism. The kidney removes H^+ and restores bicarbonate to the plasma. This is reflected in the ability of the kidney to make urine more acid than the plasma.

pH of blood 7.36–7.44: $[H^+]$ = 36–44 nmol/L
pH of urine 5.0–6.5: $[H^+]$ = 300 nmol/L – 10 μmol/L

Production of an acid urine

The following stages are involved in the production of an acid urine:

- Sodium bicarbonate is filtered at the glomerulus.
- Na^+ and some HCO_3^- is re-absorbed by the tubules.
- H^+ is secreted into the tubules after the following reaction:

$$CO_2 + H_2O \Leftrightarrow H_2CO_3 \Leftrightarrow H^+ + HCO_3^-$$
carbonic anhydrase

- HCO_3^- diffuses into the peritubular blood.
- H^+ is buffered in the tubular fluid by combining with bicarbonate, the CO_2 so formed diffuses back into the tubular cell.
- H^+ is also buffered by phosphate:
$$H^+ + HPO_4^{2-} \Leftrightarrow H_2PO_4^-$$
- H^+ is also buffered in the tubular lumen by ammonia (NH_3).

The ammonia is derived from glutamine through the action of the enzyme glutaminase. The H^+ combines with ammonia to form ammonium salts which ionise and are trapped in the tubular fluid and lost in the urine.

The amount of ammonia produced is determined by the pH ($[H^+]$) of the tubular fluid and therefore blood. The buffering capacity of ammonia can be as much as 500 mmol/day.

H^+ is secreted into the distal nephron and a gradient of 1000/L can be set up.

The minimum pH of the urine is 4.5.

Control of acid secretion by the kidney
Acid secreted by the kidney is linearly related to the P_{CO_2} of arterial blood.

Production of base
On a mixed diet the urine is normally acid.
On a vegetarian diet the urine is usually alkaline and may appear cloudy.
Base is secreted by the kidney in the form of bicarbonate.

The time-course of action of the three methods of counteracting acids entering the blood
The speed of action of the buffering systems differs
- Chemical buffer pairs – *almost instantaneously*.
- The lungs – *within minutes*.
- The kidneys – *hours to days*.

Relationship between pH, bicarbonate and P_{CO_2}
This relationship is described by the Henderson–Hasselbach equation:
$$pH = pK + \log [HCO_3^-]/\alpha P_{CO_2}$$
$$pH = pK + \log (\text{total } CO_2 - \text{dissolved } CO_2)/(\text{dissolved } CO_2)$$

Total CO_2 = 25.2 mmol/L
Dissolved CO_2 = 1.2 mmol/L
P_{CO_2} = 40 mmHg
α = 0.03 solubility coefficient
pK = 6.1
$$pH = 6.1 + \log (25.2 - 1.2)/(0.04 \times 40) = 6.1 + \log 20 = 6.1 + 1.3 = 7.4$$

$$pH \text{ (blood)} = 6.1 + \log \frac{[HCO_3^-]}{[H_2CO_3]}$$
       – controlled by the kidney
       – controlled by the lungs

APPLIED PHYSIOLOGY

When confronted with an acid–base disturbance the following questions must be answered.
1. Is there a disturbance of [H+] in the blood?
2. Is there an underlying respiratory problem? E.g. hypo- or hyperventilation.
3. Is there an underlying metabolic problem? E.g. uncontrolled diabetes.
4. Are there compensatory mechanisms operating?
   To answer these questions a number of parameters (measured on arterial blood) must be known.
- pH.
- P_{CO_2}:
   A high P_{CO_2} will occur where the lungs fail to excrete the CO_2 produced
   A low P_{CO_2} can occur if the CO_2 is lost by hyperventilation.
- Total CO_2 of the plasma.
- Base excess – the amount of alkali (negative BE) or acid (positive BE) needed to normalise blood pH, while the P_{CO_2} is held constant at 40 mmHg.

ACID–BASE DISTURBANCES

Deviations of [H+] in blood should be termed either acidaemia or alkalaemia. The terms acidosis and alkalosis should be reserved for the underlying or compensatory mechanism. The major types of acid–base disturbance include:

Metabolic acidosis

Characteristics
Low arterial pH.
Reduced plasma bicarbonate.
Compensatory hyperventilation resulting in a decrease in P_{CO_2}.

Causes
Excessive production of acid – ketoacidosis, lactic acidosis.
Excessive bicarbonate loss – gastrointestinal bicarbonate loss.
Failure to secrete H+ by the kidney – renal failure.

Respiratory acidosis

Characteristics
Low arterial pH.
Increase in plasma bicarbonate.
Elevation of P_{CO_2}.

Causes
Underventilation.

Inhibition of respiratory centre, e.g. after taking drugs.
Disorders of respiratory muscles – weak or paralysed.
Disorders of gas exchange across the alveoli.

Metabolic alkalosis

Characteristics
Elevated arterial pH.
Increased plasma bicarbonate.
Compensatory increased P_{CO_2} – due to decreased alveolar ventilation.

Causes
Loss of H^+:
   Gastric acid – vomiting
   Renal loss – excess mineralocorticoids.
Bicarbonate retention:
   Administration of $NaHCO_3$.

Respiratory alkalosis

Characteristics
Elevated arterial pH.
Low P_{CO_2}.
Reduction in plasma bicarbonate (variable).

Causes
Hyperventilation, caused by:
   High altitude
   Psychogenic factors (hysteria).
Hypoxaemia:
   Pulmonary disease – hyperventilation
   High altitude – hyperventilation.
Psychogenic hyperventilation.

10. Bone and the metabolism of calcium and phosphorus

The bones in the body make up the skeleton. This framework provides support which allows an individual to stand and move. The connections between bones, joints and skeletal muscles allow the bones to move in relation to each other. Bone is composed of cells and minerals.

BONE CELLS

Within the bone there are several cell types.

Osteocytes
This is the main cell of fully formed bone.
Probably responsible for exchange of calcium between bone and the ECF.
Plays a major role in the response to parathyroid hormone (PTH).

Osteoblasts
The cells are associated with bone formation.
Line the surfaces where bone formation is advancing.
They respond to changes in PTH and calcitonin.

Osteoclasts
The cells responsible for bone re-absorption.
Large multinucleate cells.

Fibroblasts
Found outside bone.
Relatively inactive, except after fractures.

BONE ORGANIC CONSTITUENTS AND MINERALS

Organic constituents (25%)
Collagen 90%.
Glucosamines, glycoproteins, lipids and peptides.

Minerals (75%)
Hydroxyapatite $Ca_{10}(PO_4)_6(OH)_2$. Ca/P=1.5.

Lattice structure, with Na^+, Mg^{2+}, Sr^{2+}, K^+, Cl^-, HCO_3^-, citrate, water.
Amorphous tricalcium phosphate.
The bones contain:

   99% of the body's calcium
   50% of the magnesium
   35% of the body's sodium
   9% of body water.

CALCIUM METABOLISM

99 per cent of the body's calcium is in bone, where it amounts to about 1.2 kg in a young adult. After the third decade, bone resorption exceeds accretion. This progressive loss is more pronounced in women, leading to osteoporosis.

DIETARY CALCIUM

The dietary sources of calcium include:
* Milk
* Cheese
* Green vegetables
* Artificially enriched bread.

   Average daily intake:

* Europe and the USA – 800–1000 mg.
* Developing countries – 200–400 mg.
   Recommended minimal requirements:

* Europe – 400 mg/day.
* USA – 400–800 mg/day.
* Greater in:
      Children
      Pregnancy
      Lactation.

INTESTINAL ABSORPTION

The intestine can adapt to provide an adequate intake of calcium. This occurs so that the body is always in calcium balance.

TRANSPORT MECHANISMS

The primary site for calcium absorption is the duodenum. The transport process is active and is controlled by the hormone 1,25-dihydroxycholicalciferol (1,25-DHCC). Some calcium absorption also occurs in the jejunum and ileum but here it occurs by passive

and facilitated diffusion. The proportion of calcium absorbed depends upon:
- Previous calcium intake
- Dietary constituents.
  Absorption is reduced in the presence of:
- Phytic acid
- Phosphate
- Steroids
- Excess fatty acids.
  Calcium absorption is diminished in some diseases. For example:
- Renal failure
- Intestinal malabsorption.
  Increased calcium absorption has been found in:
- Advanced age
- Associated with excess PTH and vitamin D.

PLASMA CALCIUM

The calcium concentration is maintained within narrow limits. The total plasma calcium is 2.2 to 2.6 mmol/L.
   The calcium in plasma is found in several forms:

Calcium bound to albumin	40%
Ionised calcium	45%
Complexed calcium (ultrafilterable and diffusible)	15%

   The calcium bound to protein can be affected by pH:
- Reduced in acidaemia.
- Increased in alkalaemia.

URINARY EXCRETION OF CALCIUM

Approximately 9 g/day of calcium is filtered at the glomerulus (diffusible fraction). Most of this is re-absorbed by the tubules. The urinary excretion is of the order 2–10 mmol/day (80–400 mg/day).
   Calcium re-absorption is altered in clinical conditions:
- Increased in hyperparathyroidism.
- Decreased in hypoparathyroidism.

CALCIUM IN SWEAT

15–20 mg/day (up to 100 mg/hr in extreme heat).

PHOSPHORUS METABOLISM

The total body content of phosphorus is 800 mg. This is distributed:

- Bone (80%).
- Cells (20%):
    Organic phosphates
    Phospholipids
    Nucleic acids.

PLASMA INORGANIC PHOSPHATE (FASTING)

In the plasma the total inorganic phosphate is 0.8–1.4 mmol/L (2.5–4.5 mg/100 ml). Higher values are found in infants. It is found in two forms:
1. Freely diffusible (78%)
    a. Bound to calcium and magnesium
    b. Ionised?
2. Protein bound (12%).
    At normal blood pH, 85% is present as HPO_4^{2-} and 15% as $H_2PO_4^-$.

REQUIREMENTS

Deficiencies in phosphate never occur. It is contained in all animal and plant cells.

HANDLING BY THE KIDNEY

Ninety per cent filtered is re-absorbed in the tubules.

HOMEOSTASIS OF CALCIUM AND PHOSPHORUS

The total serum calcium is controlled by the interaction between: blood, bone, kidney and small intestine. The movement of calcium between these organs is controlled by three main hormones:
1. Parathyroid hormone (PTH)
2. 1,25-Dihydroxyvitamin D or calcitriol: D [1,25-$(OH)_2$D].
3. Calcitonin.
    The rate of formation of 1,25-$(OH)_2$D is determined by the kidney. The actions of 1,25-$(OH)_2$D are to:
- Increase the rate of re-absorption of calcium from the intestine.
- Stimulate bone resorption.
- Have a role in the action of PTH on bone.

PARATHYROID GLANDS

The parathyroid glands are situated in the neck and have two major cell types:

- Chief cells secrete parathyroid hormone.
- Oxyntic cells – function unknown.

PARATHYROID HORMONE

Parathyroid hormone (PTH) conserves body calcium and increases blood calcium by increasing:
- Re-absorption from bone.
- Re-absorption from the glomerular filtrate.

Structure of PTH

PTH is a single-chain polypeptide of 84 amino acids. It originates from precursors:

$$Pre\text{-}PRH \rightarrow Pro\text{-}PTH \rightarrow PTH$$

Secretion and metabolism of PTH

The stimulus to secretion is the level of ionised plasma Ca^{2+}. As the plasma Ca^{2+} concentration falls, there is a linear increase in PTH. Secretion is not completely absent at physiological or high plasma calcium concentrations. PTH secretion is increased by β-adrenergic agents. The half-life of PTH in the plasma is short, of the order of 15 min, therefore PTH is important in the minute-to-minute regulation of secretion. It is broken down by Kupffer cells in the liver and some is removed from plasma by the kidneys.

Action of PTH

The main sites of action of PTH are kidney and bone, where it acts on both to increase plasma Ca^{2+} levels.

Actions on the kidney
- Increased calcium re-absorption.
- Decreased phosphate re-absorption.
- Hydroxylation of 25-dihydrocholecalciferol (calcidiol) to calcitriol.
- PTH actions mediated through adenylate cyclase.

Actions on bone
Increase in bone resorption – mediated by the osteoclasts.

Disorders of PTH secretion

Hyperparathyroidism
Excess production of PTH gives rise to hypercalcaemia.
Increased calcium re-absorption in the renal tubules.
Increased calcium absorption by the gut.
Increased bone resorption.

  Consequences
Patient is ill and exhibits tiredness, weakness and thirst.
Confusion, drowsiness and coma, often resulting in death.
Plasma calcium may be as high as 5.0 mmol/L.

Urinary calculi may develop.
Bone pain.
Fractures.
Renal damage.
   Causes of hypercalcaemia
Neoplasms secrete PTH.
Secondary cancer deposits in bone.
Excessive vitamin D intake.

Hypoparathyroidism
Uncommon – occurs sometimes on surgical removal of thyroid gland.
Damage to parathyroids in operations in the neck.
   Consequences
Hypocalcaemia.
Muscle spasms in hand and feet (carpopedal spasms).
Tetany.
Convulsions usually in children.
Paraesthesiae.

CALCITONIN

Calcitonin is a single-chain polypeptide of 32 amino acids. It is secreted by the parafollicular cells of the thyroid (C-cells) and is a lipophilic hormone. The main actions of calcitonin are to:
• Lower plasma calcium.
• Reduce bone resorption – inhibits osteoclasts.
• Decrease re-absorption of calcium and phosphate by renal tubules.
   Secretion of calcitonin is stimulated by:
• An increase in plasma calcium
• The polypeptide hormones:
   Gastrin
   CCK-PZ
   Glucagon.

CALCITRIOL (1,25-DIHYDROXYCHOLECALCIFEROL 1,25-DHCC)

Calcitriol is a metabolite of vitamin D produced by the renal tubules. Its output is regulated by PTH and plasma calcium.

$$
\begin{array}{l}
\qquad\qquad\qquad \text{PTH} \uparrow \text{1,25-DHCC} \\
\text{Vitamin D}_3 \longrightarrow \text{25-HCC} \\
\text{(cholecalciferol)} \qquad\qquad \downarrow \text{24,25-DHCC} \\
\text{Plasma calcium} \quad -\text{low} \rightarrow \text{1,25-DHCC} \\
\qquad\qquad\qquad\quad -\text{high} \rightarrow \text{24,25-DHCC}
\end{array}
$$

Its output is also regulated by:
   Growth hormone
   Prolactin
   Plasma phosphate.

The primary action of calcitriol is to promote active absorption of calcium by the small intestine. It is responsible for the adaptation to a low calcium diet. The processes of absorption involve:

- 1,25-DHCC binds to a protein receptor.
- Protein–receptor complex passes to the nucleus.
- Promotes the synthesis of calcium-binding protein.
- The binding protein promotes calcium absorption.
  The primary sites of action of calcitriol are:
- Bone
- Proximal renal tubules
- Parathyroid glands
- Muscle.

PHYSIOLOGICAL ROLE FOR 24,25-DHCC

When the plasma concentration of Ca^{2+} is high, 24,25-DHCC plays a part in the mineralisation of bone. Ca^{2+} is deposited and plasma Ca^{2+} is reduced.

11. Body temperature and the skin

HEAT PRODUCTION

Heat is produced in active metabolising tissues (liver, muscle, etc.) and as body tissues have a low thermal conductivity the body core would become very hot if heat were not dispersed. Heat is distributed by the blood and its transfer depends upon blood flow per mass of tissue. Heat loss is dependent upon the amount of heat delivered to the skin. Subcutaneous fat acts as an insulator whose thermal conductivity can be varied:

- Fat not perfused with blood is 0.06 $W/m^2/°C$.
- Fat perfused with blood is 0.6 $W/m^2/°C$, i.e. 10 times greater.

Note on thermal conductivity
Thermal conductivity is the flux of heat/s through 1 m^2 of material per degree Celsius per metre temperature gradient $J/s.m^2.$ $°C/m$. Since

$$1 \text{ Joule per s} = 1 \text{ watt}$$

then

$$\text{Thermal conductivity} = W/m^2/°C$$

Heat is lost from the body by physical forces. These include:
1. Conduction and evaporation from the skin. These processes are aided by air movements bringing about convection.
2. Loss from the respiratory tract $\cong$ 5% of total.
3. Radiation.

BODY TEMPERATURE

The temperature of the body varies, depending upon the site at which it is taken. The usual sites for measurement are:

- *Core or deep body temperature*, which is relatively constant.
- *Shell or superficial temperature*, which is very variable.

CORE TEMPERATURE

Sites of measurement

Mouth
This is the commonest site. Recorded values are between 35.8 and 37.3°C. The temperature may vary during the day. Mouth temperatures can lead to errors due to:
• Mouth breathing
• Drinking warm or cold fluids beforehand.

Axilla
Under the arm is frequently used. Values tend to be about 0.6°C lower than in the mouth.

Rectum
The rectum is often regarded as the best site for taking the core temperature. It is frequently used in babies. Typically values are 0.5–0.6°C higher than mouth and axillary temperatures.

Temperature of the tympanic membrane
The temperature of the tympanum is considered to be the best index by researchers.

Normal variations of body temperature
Body temperature varies during the day. This circadian rhythm can vary under different physiological conditions:
• It is highest in the morning, and lowest in the evening.
• Temperature increases slightly with exercise by about 1–2°C, depending upon intensity.
• Temperature varies during the menstrual cycle, it tends to be 0.5°C higher after ovulation.
• Temperature tends to be higher in children and lower in the elderly and the neonate.

SHELL TEMPERATURE

In temperate climates shell temperature is always higher than the environment. Therefore there is temperature gradient across the superficial tissues and a continuous loss from the skin. Skin temperatures are not uniform, being highest near the trunk.

Control of body temperature
This is achieved by balancing heat gain against heat loss.

Heat gains

Metabolic
Basal heat production (basal metabolic rate, BMR).
Muscular activity:

Muscle tone
Shivering thermogenesis – greatest from upper torso where
metabolic rate can be increased three-fold.
Specific dynamic action of food
Synthetic reactions
Brown adipose tissue (infants) – non-shivering thermogenesis.

Radiation
Heat may fall on the body as short-wave radiation from the sun or
longer wave radiation from the surroundings.

Convection
This occurs in hot climates.

Heat losses

Heat is lost by radiation and convection (70% in temperate
climates). This can be aided by vaporisation of water. This can
occur by:

* Insensible perspiration.
* Thermoregulatory sweating.
* Ventilation – through the lungs.
  Vaporisation of water can be influenced by:

* Ambient temperature.
* Relative humidity.
* Air movement.

In a cold environment

In a cold environment heat loss can be reduced by:

* Vasoconstriction.
* Behaviour:
     Clothing
     Artificial heating.

In a hot environment

In a hot environment a reduction in heat gain can be produced
by:

* Reducing the amount of clothing worn.
* Artificially cooling the environment.
* Sweating.

Thermal balance

To maintain a constant temperature it is necessary that heat gained
equals heat lost. Thermal balance is given by the following
equation:

$$M - E \pm (R + C + K) \pm S = 0$$

where:
M = rate of metabolic heat production and is always positive.
E = evaporative heat loss and is always negative.

$(R + C + K)$ = loss or gain of heat by radiation, conduction and convection. This is determined by the temperature gradient from the surface of the body to the environment.

S = change in the store of body heat. It may be positive or negative in the short term but if it does not approach zero, the time of survival is limited.

The $\pm$ attached to $(R + C + K)$ depends on circumstances, usually the environment is cooler than the body and is then negative. These can be individually calculated from physical laws, but it is easier to be done by applying Newton's Law of Cooling, i.e. 'heat flow per unit of surface area is proportional to the difference in temperature':

$$W = Km^2 (T_1 - T_2)$$

where:

W = rate of heat loss in watts.

m^2 is the area of the body.

T_1 and T_2 are the temperatures of the body (weighted mean) and environment.

K is a constant conductance.

It is more usual, when referring to the cooling of the body, to use the reciprocal of conductance, i.e. insulation I. The equation then becomes:

$$I = m^2 (T_1 - T_2)/W$$

$I = I_t + I_c + I_a$ are the insulations due to tissues of the body shell, clothing and air on the surface, respectively.

An arbitrary unit, the *clo*, is used for insulation (when designing clothes for expeditions). A clo is defined as the insulation provided by clothing sufficient to allow a person to be comfortable when sitting in a room at 21°C in still air. 1 clo represents the insulation of typical clothing of a western man.

The following table shows the total insulation of clothes plus air needed at various rates of work at low temperatures:

Activity	Metabolism W/m^2	Insulation (clo) 0°C	Insulation (clo) -20°C
Asleep	46	7	10
Resting	58	5.5	8
Light work	116	3	4.5
Moderate work	174	2	3
Heavy work	348	1	1.5

Data from Burton AC, Edholm OG 1955 Man in a cold environment. Monographs of the Physiological Society, no. 2. London: Edward Arnold.

Thermal comfort

The preferred comfort zone is an environment where a particular combination of air temperature, humidity, air movement and radiation intensity is acceptable. Most people are comfortable when

the above conditions result in a skin temperature of about 33°C, and heat balance is maintained:
- Without sweating or shivering.
- Peripheral vasoconstriction in the mid-range.

It is very variable and the elderly usually require higher background temperatures.

THERMOREGULATION

Thermoregulation is the balance between heat gain and heat loss. It is controlled by the nervous system. Body temperature is altered by:
- Small adjustments by altering skin blood flow.
- Large adjustments by shivering or sweating.

HEAT PRODUCTION

Heat production is increased by voluntary muscular effort and shivering. Muscle activity is initiated by α-motor neurones activated from the hypothalamus through tectospinal and rubrospinal tracts.

HEAT LOSS

Heat loss is controlled by the sympathetic nervous system. It is controlled from the hypothalamus. Heat loss is achieved by varying skin blood flow which alters thermal conductivity. Increased sweating activity will also increase heat loss by increasing evaporation.

REGULATION OF BODY TEMPERATURE

Changes in body temperature are detected by peripheral receptors in the skin. These respond quickly. Central receptors in the hypothalamus then respond with a time delay as a result of a warming of the blood. The core temperature is relatively slow to change because of the large heat capacity of the body. Regulation is complex, with integration between central and peripheral thermoreceptors.

Central receptors
Temperature-sensitive receptors are also found centrally in the anterior hypothalamus:
- Heat-sensitive neurones when activated cause:
    Skin vasodilatation
    Sweating and panting in animals.
- Cold-sensitive neurones (smaller in proportion) when activated cause:

Inhibition of heat-sensitive neurones
Vasoconstriction
Shivering.

The concept of an anterior heat loss centre and a posterior heat conservation centre may not be correct.

Peripheral thermoreceptors

These receptors respond to warm and cold. Recordings from afferent nerves show that both warm and cold receptors exist.

These receptors connect centrally to the:
- Cortex – conscious sensation.
- Hypothalamus.

Reflex vasoconstriction

Cold applied to a hand or a limb leads to vasoconstriction on ipsilateral and contralateral sides. The pathways involved include:

- Afferent neurone – cutaneous nerve.
- Efferent neurone – sympathetic nerves.
- Centre – spinal cord – hypothalamus.

Reflex vasodilatation

Reflex vasodilatation occurs when radiant heat is applied to the hand or body, resulting in very rapid vasodilatation. The mechanism involves:

- Afferent neurone – cutaneous nerve.
- Efferent pathway – sympathetic nerves (reduction in discharge).
- Centre – above C5 of the spinal cord.
- Reflex is inhibited if the core temperature is below 36.5°C.

Direct application of heat with water to a limb

When hot water is applied to a limb vasodilatation occurs but this has a long latent period. Its effect is due to the direct action of warm blood on the hypothalamus.

Receptors on internal surfaces

The respiratory and gastrointestinal tracts also possess thermoreceptors. Inhalation of cold air leads to shivering during inspiration. Hot food and drink can cause sweating and vasodilatation. This could also be a central effect by heating the blood.

Chemical transmission in hypothalamic centres

In animals the injection of 5-hydroxytryptamine into the anterior hypothalamus causes shivering and a rise in body temperature. The injection of noradrenaline brings about a fall in temperature. Transferring CSF from a monkey which has been cooled to another animal causes the latter to shiver. Prostaglandin E_1 injected into the lateral ventricles causes a rise in body temperature in cats and rabbits. Sodium and calcium are also important in thermoregulation.

Temperature regulation in the newborn

Fetus
Deep body temperature is 37.6–38.8°C.
Higher than mother.
Above threshold for sweating in the newborn.

At birth
1. Born wet into an environment 10°C lower than critical temperature – 32–34°C (for maintaining body temperature without increasing heat production).
2. Small body implies small heat capacity and large surface area → heat and temperature change. Heat production estimated to be 0.1 kJ/kg/min. Unclothed baby may lose 0.5 kJ/kg/min. *Therefore a negative heat balance occurs*. Thermoregulatory responses such as vasoconstriction are less effective than in adults. Shivering is negligible.
3. The principal defence against cold is *brown adipose tissue*. Activation of brown fat increases metabolic rate and heat production:
   Sympathetic activation → noradrenaline → activates adenyl cyclase → cAMP → activates lipase → oxidised in mitochondria (oxidation uncoupled) → produces only heat.
4. In the newborn exposure to heat results in vasodilatation and sweating. However, sweating is less effective than in adults.

THE SKIN

The skin covers the entire external surface of the body. Its primary functions are:

1. To act as a physical barrier between the internal systems of the body and the external environment.
2. To regulate heat loss.
3. To mediate sensation.
   The skin has three layers, the epidermis, the dermis and the subcutaneous fatty layer (Fig. 11.1).

EPIDERMIS

The epidermis is the basal layer of the skin where cells multiply. Mitotic activity varies during the day and is usually highest between 24.00 and 04.00 h. The cells which are thus formed become displaced towards the surface of the skin, forming the superficial layer. As the cells approach the surface they lose their nuclei. It takes about 2 weeks for newly formed cells to make their way to the surface of the skin. Approximately 1 g of cells are lost from the body surface per day. The epidermis can be divided into distinct layers:

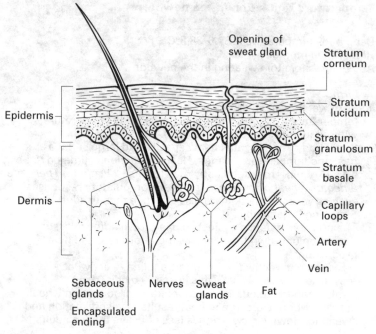

Fig. 11.1
The layers of the skin.

- Outer horny layer – stratum corneum.
- Clear layer – stratum lucidum.
- Granular layer – stratum granulosum.
- Prickle layer.
- Basal layer.

Horny layer
The horny layer has a barrier function and protects the cells of the interior against noxious agents. It also prevents water loss. There are also water-binding substances responsible for retaining water in the layer. Loss of these substances leads to drying on the horny layer and cracking of the skin. This can lead to exposure of the epithelial cells in the lower layers. The horny layer also contains free amino acids, which act as buffers against acid and alkali exposure.

Granular layer
The granular layer is responsible for the synthesis of keratin.

Prickle layer
The prickle layer contributes to the mechanical stability of the skin.

It is made up of an interlocking array of filaments attached to desmosomes, which join together the keratinocytes.

Basal layer
Projections of the basal layer extend into the dermis. This layer contains the skin pigment-producing cells – melanocytes.

Thickness of the epidermis
The epidermis varies in thickness. It is thinnest in the folds of the skin and thickest on the palms and soles of the feet, i.e. in places subjected to friction.

Skin colour
The colour of the skin depends upon the presence of pigments. These include:
- Melanin
- Oxyhaemoglobin
- Reduced haemoglobin
- Carotene.

These substances alter the reflection of light from the skin surface.

Melanin
The pigment melanin is responsible for the pigmentation of the skin at:
- The nipples
- Mammary areola
- Margin of the anus
- Sites exposed to sun and friction.

Melanin is formed by melanocytes in the basal layer of the epidermis. Melanin granules are transferred to keratinocytes by means of dendritic processes. The darker skin colour of some races is due to a greater amount of melanin and not to an increased number of melanocytes. The function of melanin is to protect the skin from injury by ultraviolet light.

DERMIS (CORIUM)

The dermis or corium is formed from a dense network of robust collagen fibres. It is responsible for the elastic properties of the skin and its ability to hold water. This is a result of the hydrophilic properties of the collagen and proteoglycans. The surface of the dermis is folded into papillae which contain:
- Blood vessels
- Lymphatics
- Nerves and nerve endings.

Capillaries do not enter the dermis.

SUBCUTANEOUS FAT LAYER

The fibrous tissue of the dermis opens out and merges with the fat-containing tissues. Collagen fibres are intermeshed with the fat cells. This layer forms a flexible link between the skin and underlying tissue and also acts as a cushion for the dermis and epidermis. An additional function is that it acts as a thermal insulator against heat loss or heat gain.

HAIR

Hairs cover a large proportion of the skin surface. This area is only slightly greater in men than women. Hairs grow from follicles. These are tubular invaginations of the epidermis. The hair is composed of a dead keratinised shaft which arises from the hair bulb. Hair can be of different types:

- *Fetal hair – lanugo hair.*
- *Birth hair – vellus hair*, which is soft and occasionally pigmented. The shafts are short and are characteristic of body hair on children and women.

Puberty – terminal hair
At puberty body hair tends to get longer, coarser and pigmented. It is sometimes called terminal hair. Terminal hair replaces vellus hair.

Hair patterns
Hair patterns can change with age and the level of androgens in the body. The distribution of hair is also genetically determined.

NAILS

Nails grow continuously out of the nail bed on the end of each digit. The rate of growth is greatest in childhood and this decreases with age. The normal rate of growth is approximately 0.5–1.2 mm/week. However, growth can be affected by general body disorders. Normally nails are not shed and are usually actively shortened. The physiological state of the individual can be reflected in the structure and appearance of the nails. The nails may become:

- Thickened
- Ridged
- Pitted
- Discoloured
- Brittle or split
- Separated from the nail bed (onycholysis).

SWEAT GLANDS

There are two types of sweat gland. These are termed the eccrine and apocrine sweat glands.

Eccrine sweat glands

These are the ordinary sweat glands. They are tubular structures found all over the skin surface. The initial part of the gland is a coiled structure in the dermis of the skin. In this region the gland is made up of a single layer of epithelial cells. There are two types of cell:

1. Dark cells, the function of which is not known.
2. Clear cells, which produce sweat.

The cells have numerous mitochondria and cannaliculi which empty into the lumen of the gland. The secretion of the gland is an isotonic solution but re-absorption can occur in the outer straight portion of the gland, the duct, as it passes out towards the skin surface.

Apocrine sweat glands

The apocrine glands are found in well-defined areas of the body. These are:

- Axilla
- Pubic region
- Areola of the breast.
  There are modified apocrine glands in the:
- Eyelid
- External auditory meatus
- The mammary gland.

Structure
The apocrine glands are branched tubes lying deep in the dermis and epidermis. They are composed of a single layer of cuboidal or columnar cell attached to a basement membrane. The ducts open into the hair follicle.

Secretions
Secretion from the apocrine glands is not activated until puberty and the secretions become reduced in the elderly. Secretion by the epithelium is continuous but the ejection of the secretion can be intermittent, resulting from myoepithelial cell activation by:

- Sympathetic nerves
- Circulating catecholamines.

The secretions are produced by the epithelial cells, axillary cells, pinching off parts of their cytoplasm and depositing this in the gland lumen. Axilliary cells secrete a viscous milky fluid. This is odourless when freshly excreted. However, it can be decomposed by bacteria on the skin surface to produce a characteristic unpleasant smell. The physiological function of this secretion is not known.

SEBACEOUS GLANDS

Sebaceous glands are made of collections of cells called acini. These cells secrete into the lumen of the acinus from where the secreted material (*sebum*) is passed into a short duct region of the gland. The ducts open into the hair follicles. However, they can open directly onto the skin surface on the face. There are no sebaceous glands on the palms, soles of the feet or the lower lip.

Secretion of sebum
Sebum is secreted by a halocrine process, i.e. secretion occurs as the result of the complete disintegration of the glandular cell.

Composition of sebum
Sebum is composed of:
- Triglycerides
- Wax esters
- Squalene
- Steroids from epithelial lipid.

Function of sebum
Sebum has the capacity to retain moisture. It protects the skin from fungal and bacterial infection. Sebaceous secretions increase markedly at and after puberty.

PROTECTIVE FUNCTIONS OF THE SKIN

The stratum corneum forms a protective barrier so that substances cannot invade the tissues and body fluids, However, some substances can penetrate the skin; the rate of penetration can vary from site to site in the body surface. Absorption can also be influenced by:

- Temperature.
- Abrasion.
- Water content of the skin (increased under a moist dressing).
- Inflammation.
- The physicochemical characteristics of the substance, e.g. solubility and molecular size.
  Substances normally pass through the horny layer via the appendages:

- Sebaceous glands
- Hair follicles.
  Some drugs can be absorbed through the skin and produce general effects. These include:

- Phenolic compounds
- Steroids.

Protection of the skin against ultraviolet (UV) light
Considerable amounts of UV light are absorbed by the horny layer due to:

- Protein

- Urocanic acid
- Melanin.
  UV light can cause damage at the cellular level. For example:

- The formation of dimers of thymidine.
- Lesions in DNA.
- Membrane effects.
  UV light can also cause damage to:

- Blood vessels – increased local production of prostaglandins.
- Collagen in the dermis.
  UV light can also affect melanin production. On exposure to UV light, melanin production increases after a latent period as a consequence of inactivation of an inhibitor of tyronase. The melanin precursor is colourless but on oxidation becomes coloured. The melanin is transferred to the keratinocytes and results in overall pigmentation of the exposed skin.

Sunburn
On exposure to sunlight, the skin becomes red (erythema). This is caused by the dilatation of papillary venous complexes. The vasodilatation can persist from 24 h to 1 week. Prolonged exposure to sunlight can result in:

- Pain
- Oedema
- Blister formation
- Systemic effects
- Vomiting
- Nausea.
  These actions are followed by pigmentation and desquamification of the horny layer.

The bacterial flora of the skin
The skin harbours small colonies of:

- Yeasts
- Bacteria
  Staphylococci
  Corynebacteria.
  The colonies are found:

- On the skin surface.
- In hair follicles.
- In sweat glands.

12. The physiology of exercise

Physical exercise is an integrated response of the energy-producing mechanisms of skeletal muscle, the respiratory, gas-transporting systems and the cardiovascular systems. Physical exercise is accompanied by increases in:

- Rate and depth of breathing.
- Heart rate and stroke volume.
- Cardiac output.
  It is usual to classify the type of exercise not only by its intensity but also by the way in which the muscles contract:

- Dynamic muscle contraction – fibres are allowed to shorten – *isotonic*.
- Static muscle contraction – fibres which do not shorten – *isometric*.

EXERCISING MUSCLE

STATIC MUSCULAR CONTRACTION

Contractions greater than 30% of maximum voluntary contraction lead to increased intramuscular tension. This compresses the blood vessels and as a result blood flow is decreased. Sustained isometric contractions give rise to:

- Sustained increase in heart rate as a consequence of vagal withdrawal.
- Relative ischaemia of muscle.
- Sustained increase in mean arterial pressure (increased systolic and diastolic).
- Small increase in left ventricular end diastolic pressure.
- Increase in cardiac contractility.
- Effects are directly proportional to the intensity of the contraction.

DYNAMIC EXERCISE

During exercise which utilises muscle shortening, i.e. walking and running, the heart rate increases as a consequence of changes in autonomic output. These changes include:

- Increase in sympathetic activity (adrenergic).

- Withdrawal of vagal activity (parasympathetic – cholinergic).
- Increase in adrenergic activity.

The increased sympathetic activity to the adrenal glands results in an increase in plasma adrenaline. At the same time there is an increase in plasma noradrenaline as a result of spill-over of noradrenaline from sympathetic junctions. This results in an increase in cardiac contractility, a noradrenergic increase in cardiac output and a rise in mean arterial blood pressure. Vasoconstriction in renal and splanchnic vascular beds causes a redistribution of blood, contributing to an increase in venous return to the heart. In the exercising muscles there is an increase in blood flow due to local metabolites. The metabolites implicated include:

- CO_2
- Adenosine
- Potassium
- Hyperosmolality.

CARDIOVASCULAR CHANGES IN EXERCISE

CARDIAC OUTPUT

See Chapter 1.

STROKE VOLUME

At rest the stroke volume is between 80 and 90 ml/beat. At maximal exercise this can increase to 110–115 ml/beat. During exercise the venous return increases with rhythmical muscle contraction. This leads to increases in intrathoracic and pulmonary capillary volumes, resulting in a pre-load to the heart. There is then an increase in pulmonary pressure causing an increased left ventricular filling. This increased end diastolic volume results in an increased force of contraction in the cardiac muscle (Starling's law of the heart) and, as a consequence, an increase in stroke volume and cardiac output. In this way there is an increased ejection fraction and a fall in end systolic volume.

HEART RATE

See Chapter X, page xx.

Control of heart rate

The heart rate increases in proportion to the workload and therefore oxygen uptake. The factors bringing about changes in heart rate come into play with different intensities of exercise:

1. From rest up to 100 beats/min – due to reduced vagal tone.
2. From 100–110 to 150–160 beats/min – due to vagal withdrawal plus an increase in sympathetic activity.

3. Above about 160 beats/min to 190–200 beats/min at $V_{O_2 max}$ – sympathetic activity.

PERIPHERAL VASCULAR RESISTANCE

The peripheral resistance is the ratio of the mean arterial pressure to the cardiac output. During exercise, the cardiac output increases but peripheral resistance falls, due to vasodilatation in muscle. The rise in cardiac output is partly achieved by an increased venous return as a result of a redistribution of blood. Thus the mean arterial pressure rises from 120/80 mmHg to 200–240/60 mmHg, i.e. systolic pressure rises and diastolic pressure falls or changes very little.

RESPIRATION IN EXERCISE

TRANSPORT OF THE RESPIRATORY GASES

The systemic arteriovenous oxygen difference under different conditions is as follows:

- *At rest:*
    Arterial O_2 content = 20 ml/dl blood or 97% saturated
    Mixed venous blood = 15 ml/dl – difference = 5 ml/dl.
- *Maximal exercise:*

    Arterial O_2 content $\cong$ 20 ml/dl
    Mixed venous blood 2–3 ml/dl – difference $\cong$ 17 ml/dl.
    The increase in the arteriovenous oxygen difference is due to an increase in oxygen extraction by skeletal muscle.

PULMONARY VENTILATION

At rest pulmonary ventilation is 10–14 L/min at a respiratory rate of 10–14 breaths/min. At maximum exercise this increases to 100–120 L/min at a rate of 49–50 breaths/min. During a period of standard exercise, P_{CO_2} and P_{O_2} change very little. The chemoreceptor threshold is reduced and body temperature is raised. The concentration of lactate in the blood increases. The ventilatory responses are also affected by:

- *Neural influences:*
    Afferent impulses from muscles and joints
    Psychological factors.
- *Venous return, increased due to:*
    Increased activity of the muscle pump
    Increased activity of thoracic pump.

Increased venous return is accompanied by pulmonary vasodilatation. This allows an increase in lung oxygen diffusion capacity which keeps the % saturation of systemic arterial blood at 97%.

The V_{O_2} max is the measure of the supply of oxygen to the muscles. The cardiac output is the limiting factor in the delivery of the V_{O_2} max, but there is a view that it is the rate of movement of O_2 from haemoglobin to the mitochondrion which may be the limiting factor.

Range of V_{O_2} max in healthy adults

Sedentary	Normally active	In training	Endurance athletes
30 ml/kg/min	45 ml/kg/min	53 ml/kg/min	80 ml/kg/min

Although the mean capillary transit time speeds up in exercise, it is still sufficient time to bring about alveolar–oxygen equilibration.

MUSCLES WITH A FREE SUPPLY OF OXYGEN

With a light or moderate workload the immediate source of energy is ATP. The concentration of ATP in muscle $\cong$ 7–10 mmol/kg muscle wet weight. This store is sufficient for only for about 2 min of moderate exercise. As a result, ATP has therefore to be continuously regenerated. ATP is re-synthesised from:

- Creatine phosphate.
- Glycogen and fat in muscles.
- Glucose and free fatty acids from blood.
  The RQ (respiratory quotient) establishes the relative proportions of substrate used. In exercise, the blood glucose concentrations remain constant. However, in prolonged exercise:

- Glycogen stores run down.
- Relatively more fat is used.
- Glucose passes from blood to muscles.
  Liver glycogen is limited, despite an increase in gluconeogenesis (lactate and protein).

Oxygen utilisation by muscles
The oxygen utilisation by muscles depends upon circulatory adjustments. These include:

- Vasodilatation.
- More capillaries perfused.
- Diffusion distance to muscle fibres decreased.
  Blood is redistributed from the:

- Kidneys.
- Abdominal viscera.
- Skin – later vasodilatation occurs to promote heat regulation.

Oxygen transport system
The % extraction of oxygen from the blood is determined by:

- Po_2 at the mitochondria.
- High Pco_2.
- High $[H^+]$.
- 2,3-Biphosphoglycerate.
  A-V difference is greater in exercise and in trained athletes.

MUSCLES WITH A LIMITED SUPPLY OF OXYGEN

During short-duration exercise the metabolic work is greater than predicted from the O_2 consumption. This metabolic need is met from anaerobic metabolism. The workload is limited by:

- Exhaustion of energy supplies.
- Accumulation of lactic acid.
  Energy needed still comes from ATP which, under anaerobic conditions, is synthesised from:
- Phosphocreatine
- Glycogen.
  These processes lead to an accumulation of lactate which diffuses into the blood and contributes to fatigue. The lactate is subsequently converted to glucose and glycogen in liver (gluconeogenesis: see Ch. four).
  After exercise is over the O_2 consumption remains elevated and returns only slowly to control levels. This is termed the oxygen debt of exercise. In the repayment of the debt, for every 2 moles of lactate produced 1.7 moles are re-synthesised in the liver, and 0.3 moles oxidised to CO_2 and H_2O.

BLOOD GLUCOSE DURING EXERCISE

During exercise blood glucose is constant unless the exercise is prolonged. In normal exercise the intake of glucose by the muscles is increased and there is an increased entry of glucose into blood from the liver (hepatic glycogenolysis). Glucose is derived through gluconeogenesis in long-term exercise.

NEUROHUMORAL CHANGES IN EXERCISE

The supply of substrate for energy production is under neurohumoral control. In exercise, there is an increase in sympathetic activity which influences:

- Glyogenolysis.
- Glucagon secretion.
- Depression of insulin secretion.
- Increase in plasma free fatty acids.
  Increase in secretion of other hormones associated with growth and repair of tissue:

- Growth hormone – mobilises fat.
- 17-Hydroxysteroids from adrenal cortex.

PHYSIOLOGICAL INDICES OF PERFORMANCE

The two main physiological indicators of performance are:

- Maximum oxygen consumption.
- Anaerobic threshold.

13. Membranes and signals

MEMBRANE STRUCTURE AND FUNCTION

PLASMA MEMBRANE

Plasma membrane or cell membrane forms a barrier between the cell *cytoplasm* and the *extracellular fluid* (ECF).

Composition
Phospholipids
Proteins
Glycoproteins.

Structure
Lipid bilayer – hydrophobic 'heads' outermost, lipophilic 'tails' in the middle of the membrane.
Integral proteins – either spanning across the bilayer or facing the cytoplasm or ECF.

Functions
Separates cytoplasm and ECF.
Maintains cell's internal environment.
Transports macromolecules into and out of the cell.
Controls Na^+, K^+, Ca^{2+} and Cl^- ion distributions between cytoplasm and ECF.
High electrical resistance.
Generates trans membrane voltage difference – the membrane potential.
Protein receptors for hormones and transmitter substances.
Signal transduction.

INTRACELLULAR MEMBRANES

Intracellular membranes form boundaries between intracellular organelles and compartments and the cell cytoplasm.

Nuclear membranes
Surround the nucleus.
Influence mitosis.
Protein synthesis and modification.

Sarcoplasmic and endoplasmic reticulum (SR/ER)
SR found in skeletal, cardiac and smooth muscle.
ER similar to SR but in non-muscle cells.
SR functions as a Ca^{2+} store used in the activation of muscle contraction.
Ca^{2+} is taken into SR by a pump utilising ATP.
The ER is subdivided into rough and smooth ER (RER/SER).
RER is distinguished from SER by the appearance in the electron microscope of small dense particles (10–20 nm) called ribosomes. Ribosomes consist of protein and RNA; their function is to synthesise protein.
SER is a:
    Ca^{2+} store
    Site of hormone and glycoprotein synthesis and modification.

Lysosomes
Contain hydrolytic enzymes.
Discharged into vacuoles to break down proteins, nucleic acids and foreign particles, e.g. bacteria.

Golgi apparatus
Packages proteins made in RER.
Particularly important in secretory cells.

Mitochondria
Each mitochondrion has two membranes:
    Outer membrane – lipid bilayer containing proteins
    Inner membrane – forming folds or cristae, site of oxidative phosphorylation and ATP production.

TRANSPORT ACROSS THE CELL MEMBRANE

NON-CHARGED MOLECULES – HYDROPHOBIC

The lipid bilayer is a barrier to the passage of all molecules into and out of the cell.
Non-charged substances and gases move freely across the membrane by dissolving in the lipid and diffusing down their concentration gradient.
The rate of diffusion depends on the lipid solubility and concentration gradient.
Water moves freely across the membrane.

CHARGED MOLECULES – HYDROPHILIC

The lipid bilayer makes it difficult for charged molecules to enter or leave the cell. The cell uses specialised proteins (channels, ion exchangers, cotransporters and pumps) to allow transport.

The movement of such molecules is affected by their charge, the membrane potential and the concentration difference across the membrane (electrochemical concentration gradient).

FACILITATED DIFFUSION

Molecules move down their electrochemical concentration gradient. Separate mechanisms exist for different molecules, although similar molecules may be transported by the same mechanism. This may lead to competition between them for transport. The number of transport sites is finite, therefore there is an upper limit to the amount of substance that can be transported (saturation).

Examples include:

- Ion channels – responsible for the membrane potential and action potential.
- Sugar transporters – the entry of glucose.
- Amino acid transporters.

PRIMARY ACTIVE TRANSPORT

Molecules can be transported against their electrochemical gradient. Such processes require energy to be expended (the hydrolysis of ATP to ADP) and are thus active. Mechanisms can exchange molecules or operate on single molecules.

Examples include:

- Na^+–K^+ exchange pump – maintains high intracellular K^+ and low intracellular Na^+.
- Plasma membrane Ca^{2+} pump – extrudes Ca^{2+} out of the cell.
- ER/SR Ca^{2+} pump – accumulates Ca^{2+} in the ER/SR.

SECONDARY ACTIVE TRANSPORT – Na^+-COUPLED COTRANSPORT

These mechanisms do not use energy but use the Na^+ concentration gradient – Na^+ gradient maintained by the Na^+–K^+ exchange pump.
Molecules can enter the cell with Na^+.
Molecules can exit the cell in exchange for Na^+ entering.

Examples include:

- Na^+/H^+ exchange – regulates intracellular pH.
- Na^+/Ca^{2+} exchange – regulates intracellular Ca^{2+}.
- Na^+/K^+/Cl^- cotransport – carries K^+ and Cl^- into the cell.
- Na^+–glucose transport – carries glucose into the cell.
- Na^+–amino acid transport – carries amino acids into the cell.

ENDOCYTOSIS AND EXOCYTOSIS – MEMBRANE TURNOVER

Endocytosis
Large molecules and bulk movement of material are carried into the cell by capturing sections of the surface membrane and external environment to form coated vesicles. Examples include:

• Insulin uptake.
• Macromolecule transfer across capillaries.

Exocytosis
Molecules, synthesised within the cell, are packaged into membrane-bound vesicles, incorporated into the surface membrane and so exit the cell. Examples include:

• Secretion of hormones.
• Neurotransmitter release.

THE INTERNAL ENVIRONMENT OF THE CELL

These processes, active membrane pumps, Na^+-coupled transporters and exchangers, function in unison to maintain the internal composition of the cell constant different from the external environment. These differences are necessary to keep the cells alive and are used to carry out specific cell functions.
• Resting intracellular K^+, 140 mM; Na^+, 10 mM; Ca^{2+}, 10^{-7} M.
• Resting pH of the cytoplasm is pH 7.2.
• A high intracellular Ca^{2+} (mM) kills cells.
• Alterations in pH (H^+ concentration) will make enzymes less efficient.
• Na^+, K^+ and Ca^{2+} gradients need to be maintained to generate electrical activity.
• The Ca^{2+} gradient is used to activate secretion and contraction.
• Cotransporters used to move ions and water across whole cells (epithelia).

RESTING MEMBRANE POTENTIAL AND ION FLOW

An electrical potential exists between the inside of a cell and the outside – the membrane potential.
   The cell maintains a different ionic environment inside compared to outside, as a result of membrane pumps and transporters. Specialised proteins in the lipid bilayer (ion channels) allow the movement of selected ions across the membrane. This movement gives rise to current flow and potential changes.

IONIC BASIS OF THE RESTING POTENTIAL

- The difference in K^+ concentration between the outside and inside of the cell is set up by the Na^+-K^+ exchange pump.
- External K^+ – 5 mM, internal K^+ – 140 mM.
- At rest, the majority of channels operating are K^+ channels which allow K^+ to exit the cell.
- The resulting loss of positive charge generates a negative potential within the cell.
- This process proceeds to an equilibrium where K^+ loss is balanced by the internal negativity – there is no net movement of K^+.
- This is the potassium equilibrium potential for potassium (E_k) and can be calculated from the Nernst Equation:

$$E_k = (RT/zF).\ln [K]_o/[K]_i$$

where:
R is the gas constant
T is temperature
z is the valency of the ionic species
F is Faraday's Constant
$[K]_o$ is the extracellular concentration of potassium
$[K]_i$ is the intracellular concentration of potassium.

The membranes of many cells at rest allow other ions to move across, e.g. Na^+ and Cl^-. Under these conditions the resting membrane potential is influenced by the transmembrane concentration gradients of these ions and the number of ion channels allowing them to pass. The membrane potential (E_m) is then derived using a modified Nernst Equation, the Goldman Equation:

$$E_m = (RT/zF).\ln (P_k[K^+]_o/P_k[K^+]_i + P_{Na}[Na^+]_o/ P_{Na}[Na^+]_i + P_{Cl}[Cl^-]_i/ P_{Cl}[Cl^-]_o)$$

P is the membrane permeability for each ion and is related to the number of ion channels and the ease of movement of each ion across the membrane. When:

$$P_k >> P_{Na} > P_{Cl}$$

then:

$$E_m = (RT/zF).\ln [K^+]_o/[K^+]_i$$

The Na^+-K^+ exchange pump extrudes 3 Na^+ for every 2 K^+ transported into the cell. Thus when active, there is a net movement of positive charge out of the cell, thereby increasing the intracellular negativity. The Na^+-K^+ exchange pump is said to be *electrogenic*.

ION CHANNELS

Ion channels consist of specialised proteins in the cell membrane whose function is to facilitate the passage of ions into or out of the cell or between cells. They are found in all cells in the body.

TYPES OF ION CHANNEL

Voltage-operated channels (VOCs)

Distribution
Found in all excitable cells (nerve, skeletal muscle, smooth muscle).

Structure
Channels are proteins with four subunits all the same (homotetramer). Each subunit is a polypeptide which spans the membrane several times. Elements on the outer or inner aspect act as voltage-sensitive gates to open or close the channel. Drugs and poisons bind to channels to interfere with opening or closing.

Function
Different channels are selective for different ions (Na^+, K^+ and Ca^{2+} channels).
These channels underlie the electrical activity of cells, e.g. the action potential.

Regulation
VOCs can be regulated by phosphorylation and G-protein interaction.

Receptor-operated channels (ROCs)

Distribution
Found on almost all cells in the body, mainly at synapses.

Structure
Channels consist, typically, of five subunits of which several are different. Each subunit is a polypeptide spanning the membrane several times. Specific subunits perform particular functions, e.g. ligand binding. The majority are not specific for cations and allow Na^+, K^+ and Ca^{2+} to pass. Some ROCs specifically allow anions to pass, e.g. Cl^-.

Function
ROCs open as the result of a specific molecule (ligand) binding to a complementary structure (receptor) on the protein. Their function is to alter the membrane potential resulting in a depolarisation or hyperpolarisation depending on the ion passing through.

Regulation
ROCs can be regulated by phosphorylation and G-protein interaction.

Gap junctions

Distribution
Gap junctions are specialised channels which allow cell-to-cell communication.
Found in cardiac muscle, smooth muscle and epithelia but not in neurones (in higher species) or skeletal muscle.

Structure
Highly non-selective channels allowing all ions and macromolecules (molecular weight up to 1000) to pass. On each cell the junction consists of six similar subunits. Each set of subunits is aligned to form a common pore.

Function
In cardiac and smooth muscle their function is to provide a low resistance path for the spread of ionic current. In other cells they facilitate the passage of metabolites and intracellular signals between cells.

Regulation
Regulated by Ca^{2+}, H^+ and transcellular voltage.

SIGNALLING MECHANISMS

There are a variety of different mechanisms on the cell surface which detect substances outside the cell and lead to the generation of an intracellular signal. These include:

Receptor-operated ion channels
These structures are also known as transmitter-gated ion channels and are found at synapses. They have been described on page 354.

G-protein linked systems
These are the largest family of cell surface receptors. The receptor proteins recognise neurotransmitter substances, hormones and local factors. This system involves a receptor, an intermediary membrane protein (G-protein) to a target protein.

Receptors
The receptors are a family of proteins which are a single strand of protein which spans the membrane seven times. The sites which interact with the G-proteins are located on the cytoplasmic face of the protein. The internal sections can be phosphorylated and this process can lead to desensitisation of the receptor.

G-PROTEINS

There are two families of linking proteins – *single strand monomeric proteins* and *trimeric GTP-binding regulatory proteins*. Both monomeric and trimeric G-proteins are enzymes which bind guanosine triphosphate (GTP) and in this state are active. In the active state the G-protein can activate many different target proteins. Cleavage of the bound GTP to GDP inactivates the protein and prevents its positive action on the target protein.

SEQUENCE OF EVENTS

- Ligand binds to the extracellular domain of the receptor.
- GTP is ejected from the inner surface and is replaced with GTP.
- The G-protein is now active and is capable of activating the target protein.
- Bound GTP is cleaved to GDP and the G-protein is now inactive and does not activate the target protein.

TARGET PROTEINS

The target protein can be an enzyme or an ion channel.

EXAMPLES OF G-PROTEIN-LINKED SIGNALLING SYSTEMS

1. G-protein linking to adenylate cyclase and cellular cAMP levels

G_s-proteins
Adenylate cyclase is a membrane-bound enzyme which catalyses the formation of cAMP from ATP. Cellular levels of cAMP are known to rise in response to hormone and extracellular signals. The link between ligand binding and cAMP production involves G-protein activation of adenylate cyclase. Because the G-protein is stimulatory to adenylate cyclase, it is called G_s.

- G_s is a trimeric G-protein and is composed of three distinct polypeptide chains, α_s, β and γ.
- α_s binds GTP, activates adenylate cyclase and hydrolyses GTP to GDP.
- The b and c chains form a tight complex (β/γ) which functions to hold the entire α_s–β/γ complex to the inner surface of the membrane.
- The α_s–β/γ G-protein complex has bound GTP and is inactive.
- When the receptor is activated by ligand binding and the receptor associates with the α_s–β/γ G-protein complex, GDP is exchanged for GTP.
- α_s then dissociates from the b/c complex and associates with adenylate cyclase.

- Active adenylate cyclase leads to the production of cAMP.
- α_s bound to GTP is a short-lived molecule and GTP is rapidly hydrolysed to GDP.
- α_s-GDP is now inactivated and cannot activate adenylate cyclase.
- α_s-GDP re-associates with the b/c subunits and can interact with another activated receptor if it is available.
- Activated receptors will remain available as long as the ligand is present in the extracellular space.

Noradrenaline Adrenergic stimulation can lead to the production of cAMP. The receptor involved in noradrenaline binding is b and this activates Gs.

Cholera toxin Cholera toxin is a bacterial toxin which has profound effects on the gastrointestinal tract. Cholera toxin is an enzyme which leads to a modification of the α_s subunit, such that it cannot hydrolyse GTP. Thus the activated G-protein cannot be inactivated and cAMP production proceeds without check. The cellular consequences of an elevated cAMP can be serious. In the gut it can cause severe diarrhoea.

G_i-proteins Some hormones are known to decrease cellular levels of cAMP. The link between ligand binding and cAMP production involves a G-protein whose activation leads to an inhibition of adenylate cyclase. Because this G-protein is inhibitory, it is called G_i.

Noradrenaline
Adrenergic stimulation can also lead to a decrease in the production of cAMP. The receptor involved in the noradrenaline binding and action on these occasions is α_2. Receptor stimulation leads to the activation of an inhibitory G-protein – G_i.
- Activation of α_2 receptors leads to the activation of the G_i trimeric complex.
- The α_i subunit releases its GTP and binds GTP.
- The activated G_i complex dissociates into α_i and β/γ.
- α_i inhibits adenylate cyclase, as does the β/γ complex.
- β/γ also binds to α_s subunits, causing them to become inactive.
- α_i hydrolyses GTP and the α_i protein re-complexes with β/γ.
- the α_i–β/γ complex can be activated again if the α_2 receptors are still occupied.

Cellular actions of cAMP
cAMP is a well-recognised intracellular signal responsible for activation or modulation of a large number of intracellular systems. Its main pathway for modulation involves the activation of *cAMP-dependent protein kinase (A-kinase)*.
- A-kinase is present in all cells.
- The substrates for A-kinase differ from cell to cell, allowing cAMP to activate or modulate many different cellular processes.

- Inactive A-kinase consists of four subunits, two regulatory and two catalytic subunits.
- If the cytoplasmic concentration of cAMP increases, two molecules of cAMP bind to each of the two regulatory subunits.
- Once this has happened the two catalytic subunits dissociate and become two separated active catalytic units.
- The active units can then activate phosphorylating enzymes.

A-kinase is an important enzyme in the production of a variety of signalling molecules generally known as the *eicosanoids*. These molecules are made primarily from precursors, mainly arachidonic acid, which is derived from membrane phospholipids by the action of phospholipases. There are four main types of eicosanoid:

- Prostaglandins
- Prostacyclins
- Thromboxanes
- Leukotrienes.

The synthesis of the prostaglandins, prostacyclins and thromboxanes involves the enzyme cyclo-oxygenase, while the synthesis of leukotrienes involves the enzyme lipoxygenase (B).

2. G-protein linking to phospholipase C

Phospholipase C-β (PLC-β) is a member of a family of membrane-bound enzymes found in almost all cells. The function of PLC-β is to phosphorylate the membrane lipid phosphotidyl bis phosphate (PIP$_2$) to produce *inositol trisphosphate (IP$_3$)*, a water-soluble molecule which rapidly distributes throughout the cell, and diacylglycerol, a lipid derivative which can activate a Ca^{2+}-dependent enzyme – *protein kinase C (PKC)*, which can phosphorylate a variety of intracellular proteins.

Sequence of events on PLC-β activation
Activated receptor proteins in the cell membrane activate a tetrameric G-protein called G$_q$.
Activation of G$_q$ releases GDP and dissociates the α_q subunit from the b/c subunit complex.
The G$_q$ subunit binds GTP and is now in an active state. In this state G$_q$ activates PLC-β.
Active PLC-β converts PIP$_2$ to IP$_3$ and diacylglycerol.

Cellular actions of IP$_3$
IP$_3$ is a small molecule whose primary action is to act on specific proteins on the internal membranes of cells – *IP$_3$ receptor channel complexes*. The IP$_3$ receptor channel complexes function as IP$_3$-gated Ca^{2+} channels. In the presence of IP$_3$, the channels are opened and Ca^{2+} is released from the intracellular stores. This Ca^{2+} is then used for activating many cellular processes. The IP$_3$ receptor channel complex has four identical subunits with the central region forming the Ca^{2+} channel. The complex can be made more sensitive

to IP_3 when the cytoplasmic Ca^{2+} concentration is elevated just above basal levels and can be inhibited at high Ca^{2+} concentrations.

ENZYME-LINKED SYSTEMS

In this system, binding of a ligand to the external face of the membrane receptor leads to the formation of an active catalytic domain at the cytoplasmic face of the molecule. These systems are typically protein kinases or are associated with protein kinases which lead to phosphorylation of intracellular proteins or enzymes.

14. Skeletal muscle and its properties

STRUCTURE – GENERAL

MOTOR UNITS

Skeletal muscle consists of large numbers of elongated cells.
Tendons at each end attach muscles to bone.
Muscle cells are referred to as *muscle fibres*.
Fibres have a striped appearance – *striated muscle*.
Each muscle fibre is innervated by a single motor nerve.
One motor nerve may innervate many muscle fibres – the *motor unit*.
When a single motor nerve fires the motor unit contracts as a single unit.
Number of fibres in a motor unit:
   10 for extraocular muscle – fine movements
   2000 for calf muscle – coarse movements.

STRUCTURE – CELLULAR

MUSCLE FIBRES

Membranes
The plasma membrane of muscle is called the *sarcolemal membrane*.
A system of fine tubes extends from the sarcolema deep into the muscle fibre – the *T-tubule system*.
A second system of fine tubules runs longitudinally along the fibre – the *sarcoplasmic reticulum* (*SR*).
The T-tubules are not directly connected to the SR.
SR contains a Ca^{2+} binding protein.
Where the T-tubules meet the SR there are specialized contacts – *triads*.

Organelles
Each fibre contains many *nuclei* (multinucleate) and *mitochondria*.

Contractile apparatus

The contractile elements are arranged in units in the fibre –
myofibrils.
Myofibrils consist of the contractile proteins *myosin* and *actin*.

Myosin
Myosin is the *thick* filament and has two parts:
    Tail – two coiled peptide chains (light meromyosin).
    Head – two units of heavy meromyosin each having two light
    chains.
    Light chains are associated with the ATPase activity.

Actin
Actin is the *thin* filament and consists of two helical strands of
protein (filamentous actin).
Each strand is made of identical subunits of globular actin.

Tropomyosin
Tropomyosin is a protein which lies in the two grooves of the actin
helix.
Troponin is a protein bound to tropomyosin at regular intervals.
Troponin molecules are spaced 1 every 7 G-actin monomers.
Troponin has three components:
    T – interacts with tropomyosin
    I – inhibits binding of myosin ATPase
    C – Ca^{2+} binding site.

Striated pattern

A bands – myosin filaments and interdigitating actin filaments.
I bands – area of actin not overlapping with myosin.
H zone – area of myosin not overlapping with actin.
Z lines – anchor points for actin filaments.

Elastic components

The muscle structures not associated with the contractile apparatus
affect its mechanical properties.
Non-active muscle behaves like an elastic spring.
For small increases in length little force needs to be applied.
As the length increases greater force is needed to make the same
change in length – *Hook's law*.

INNERVATION

α motor neurone – fast conduction – 70 m/s.
Motor neurones are myelinated except for the final terminal region
in the junction.

Neuromuscular junction (NMJ)

Presynaptic terminal
Surface membrane has VOC Ca^{2+} channels.
In the cytoplasm there are numerous membrane-bound vesicles containing transmitter substance – acetylcholine (ACh).
Mitochondria.

Synaptic cleft
Gap of 70–100 nm.
Contains basement membrane.
Acetylcholinesterase (acetylcholine-destroying enzyme) found in cleft.

Postsynaptic membrane – end plate
Specialised region on muscle fibre.
Groove on fibre surface in which motor nerve sits.
Membrane highly folded – 2000 μm^2.

Neuromuscular transmission
Na^+-dependent action potential invades nerve terminal.
VOC Ca^{2+} channels activate and initiate Ca^{2+} influx.
Vesicles migrate to presynaptic membrane.
Vesicles dock to specific proteins.
Vesicle fusion and discharge of ACh into cleft – exocytosis.
200–300 vesicles involved per nerve impulse.
ACh diffuses across cleft.
Partial inactivation by acetylcholinesterase.

ACh binds to receptors on junctional folds.
Opening of ACh channels (ROC).
Increased flux of Na^+ and K^+.
Depolarisation of the muscle end plate – end plate potential.
Adjacent muscle membrane depolarises by current spread.
Muscle Na^+-dependent action potential initiated.
The initiation of excitation-contraction coupling.
Residual ACh destroyed by acetylcholinesterase.

One action potential in motor neurone gives rise to one action potential in the muscle fibre.

Acetylcholine receptor and ion channel at the NMJ
Activated pharmacologically by nicotine – nicotinic receptor.
The receptor/channel complex is a structure made up of five subunits – 2α, β, γ and δ – each spanning the lipid bilayer.
The subunits link to form an aqueous channel.
Each α subunit binds one ACh molecule.
When two ACh molecules bind the channel opens.
Both Na^+ and K^+ pass through the channel.

End-plate potential (EPP)
The depolarisation produced by activation of the ACh receptor.
4000 molecules of ACh contribute to each EPP.
Depolarisation is the result of simultaneous Na^+ and K^+ current flow.
Located only in the region of the synaptic cleft.
Timecourse depends on the destruction of ACh.
Every EPP depolarises adjacent membrane to threshold and initiates muscle action potential.

EXCITATION-CONTRACTION COUPLING

Once the muscle action potential (AP) is initiated there follows a sequence of events which result in contraction of the muscle fibre – excitation-contraction coupling.

Excitation
Initiation of the muscle AP at end plate.
AP spreads along sarcolema and down into t-tubules.
Depolarisation of t-tubules is sensed by the lateral SR at the triads.
Muscle AP lasts for 5 ms.
Ca^{2+} is released from terminal SR.
Ca^{2+} release activates more Ca^{2+} release (Ca^{2+}-induced Ca^{2+} release).
Cytoplasmic Ca^{2+} rises.
There is a delay between the AP and sufficient Ca^{2+} to activate contraction.
In response to one AP Ca^{2+} rises to 20–30 µM.
Ca^{2+} remains elevated for 50–60 ms.

Cross bridge cycle
Ca^{2+} binds to troponin C.
Conformational change in tropomyosin.
Myosin binding site revealed.
Myosin head attaches to tropomyosin.
Myosin head.
Actin and myosin filaments slide past each other – *power stroke*.
ATP binds to myosin head group.
Myosin head detaches from actin.
ATP is hydrolysed to attached ADP and free phosphate.
Myosin head is now reprimed for attachment.

Contraction
In the presence of Ca^{2+} and ATP cross bridge cycle continues.
Fibre length decreases as filaments slide past each other.
Contraction continues as long as Ca^{2+} is elevated.

Twitch duration is approximately 100 ms.
Elevated cytoplasmic Ca^{2+} activates SR Ca^{2+} pump.
Cytoplasmic Ca^{2+} falls.
Crossbridge cycles ceases.
Muscle stops contracting.

Contraction resulting from actin and myosin movements is called the *sliding filament theory*.

FUNCTIONAL PROPERTIES OF MUSCLE

Twitch and tetanus
The contraction is response to one AP is called a *twitch*. A second AP arriving after the first will activate a second twitch. If the time between the AP is short (10 ms) then the forces generated by each AP add together. During a train of impulses twitches become fused into a smooth contraction – *tetanus*.

Isometric and isotonic contractions – experimental conditions
When a muscle is activated to produce force but not allowed to shorten, this is isometric contraction. When a muscle is allowed to shorten while generating a constant force, this is isotonic contraction.

Length – tension relationship – isometric twitch
As muscle is stretched the force generated by a twitch increases. There is an optimum length to produce maximum force (L_{max}). At lengths > L_{max} force production declines; this is the *length–tension relationship* of striated muscle. The length–tension relationship reflects the degree of functional overlap between actin and the number of active myosin head groups. As the length increases more head groups overlap and contribute to force production.

At L_{max} the maximum number of head groups are operational. Beyond L_{max} the myosin is pulled past the actin filaments and the number of head groups able to attach decreases. Under these circumstances force production decreases.

Force–velocity relationship – isometric twitch
As the muscle shortens myosin heads move along the actin filaments. During this movement only a fraction of the crossbridges are attached and creating force. An isometric twitch therefore produces less force than a isotonic twitch. For a low load only a few crossbridges need to be attached to generate enough tension to move the load. This allows a rapid shortening. Maximum velocity is achieved at no load.

For higher loads more crossbridges need to be active at any one time therefore shortening is slowed. If insufficient force is generated to move the load then the contraction remains isometric.

The relationship between speed of shortening and force generated is the *force–velocity relationship*.

Muscle stiffness
Muscle stiffness is a property of the crossbridges. An additional load on a steadily contracting muscle causes additional tension to be developed proportional to the active tension and the velocity of stretch. These events resist extension.

Contractions within the body
When the contractile apparatus is activated the initial force is transmitted to the elastic components of the fibres. Only a fraction of the force produced is transmitted to the load. When the tension in the elastic components is sufficient to move the load, the muscle shortens and no increase in tension occurs. To move a load, the contraction is first isometric then isotonic.

Voluntary control of muscle contraction
Force is modified in two ways:
   Number of active motor units
   Rate of activation of units.
Motor unit firing is asynchronous giving a smooth whole muscle contraction. Muscles with small motor units are used for fine movements in motor control.

Denervation of muscle
Destruction of the nerve to a muscle leads to:
   Atrophy of fibres
   Degeneration of the nerve
   Muscle fibrillation:
      Endplates become diffuse
      ACh sensitivity all over fibre
      Generate spontaneous action potentials
      AP smaller and prolonged
      Na^+ channels are tetrodotoxin resistant.
Without the nerve muscle is irreversibly lost.
Trophic factors from nerve are essential for muscle survival.

Trophic effects of nerves
Re-innervation of a fast muscle with a nerve originally to a slow muscle leads to:
   A slower twitch
   Changes enzyme profile to a slow fibre type
   New capillaries appear
   Resting blood flow increases.

The opposite changes occur if a slow muscle is re-innervated by a nerve originally innervating a fast muscle.

Effects of physical training
Training leads to muscle hypertrophy by the creation of new muscle fibres.
Fibres increase their metabolic enzymes.
New capillaries form.

MUSCLE FIBRE TYPES

There are in general two types of muscle fibre:
Fast – phasic/twitch muscle
Slow – postural/tonic muscle.

	Fast muscle	Slow muscle
Mechanical properties		
Speed of contraction	Very variable	Very variable
Velocity of isotonic shortening	Faster	
Rate of relaxation	Faster	
Rate of tension rise	Faster	
Rate of relaxation	Faster	
Tetanus fusion frequency	Higher	
Work load		More effective for a given work load
Electrical properties		
Action potential	Shorter	
Propagation velocity	Faster	
End plates	More sensitive to competitive inhibitors	More sensitive to depolarisation block
Biochemical properties		
Myoglobin content	Low	High
Oxidative enzymes	Low	High
Fatigue	Rapid	Resistant
Glycogen content	High	Low
Glycolytic enzymes	High	Low
Sarcoplasmic reticulum	Sequester Ca^{2+} more rapidly	
Capillary density		Greater
Resting blood flow		Greater

SENSORY INNERVATION OF MUSCLE

Nerve endings:
Pain.
Golgi tendon organs.
Muscle spindles.

PAIN ENDINGS

Signal fatigue and damage.

GOLGI TENDON ORGANS

Mounted in tendons in series with muscle fibres.
Connected to 1B nerves.
Frequency of discharge rises with increased tension.
During active movements their discharge has a reciprocal
relationship to that of muscle spindles.

MUSCLE SPINDLES

Receptors are in parallel with extrafusal fibres.
Intrafusal fibres lying within a fluid-filled capsule, poles attached to
connective tissue, cross-striated and contractile, are of two types:
  Nuclear bag fibres – nuclei bunched in middle causing a bulge.
  Nuclear chain fibres – thinner with nuclei lined up in a row.
Afferent innervation:
  Primary fibres (Group 1A) – annular spiral ending. Discharge
  dynamic, show adaptation, respond in proportion to rate of
  change in velocity.
  Secondary fibres (Group II) – partly annulo-spiral and partly
  flower spray endings on nuclear chain. Discharge proportional to
  degree of stretch – non-adapting (static).
Efferent innervation:
  Spindles receive a motor innervation from γ-fibres; which cause
  contraction of poles and increase sensitivity of spindle to stretch.
  γ_s – static – to nuclear chain – increase sensitivity to both of both
  secondary and primary fibres to static stretch.
  γ_d – dynamic – to nuclear bag – increase sensitivity to group 1A
  fibres but not to group II fibres.

CENTRAL CONNECTIONS

Fibres from Golgi tendon organs and muscle spindles:
Project through a number of pathways to the cerebellum
(proprioceptive information).
Involved in reflex effects within the spinal cord.

Index